MANAGEMENT TECHNIQUES
APPLIED TO
THE CONSTRUCTION INDUSTRY

R. OXLEY,
M.Sc., M.I.O.B., A.M.I.B.M.

and

J. POSKITT,
M.I.O.B., A.M.B.I.M.

CROSBY LOCKWOOD STAPLES
GRANADA PUBLISHING
London Toronto Sydney New York

First published in Great Britain 1968 by Crosby Lockwood & Son Ltd

Second edition 1971

Reprinted by Granada Publishing Limited
in Crosby Lockwood Staples 1976, 1979

Granada Publishing Limited
Frogmore, St Albans, Herts AL2 2NF
and
3 Upper James Street, London W1R 4BP
1221 Avenue of the Americas, New York, NY 10020, USA
117 York Street, Sydney, NSW 2000, Australia
100 Skyway Avenue, Toronto, Ontario, Canada M9W 3A6
110 Northpark Centre, 2193 Johannesburg, South Africa
CML Centre, Queen & Wyndham Streets, Auckland 1, New Zealand

Copyright © 1971 by R. Oxley and J. Poskitt

ISBN 0 258 96847 8

Printed in Great Britain by
Spottiswoode Ballantyne Ltd.
Colchester and London

MANAGEMENT TECHNIQUES APPLIED TO
THE CONSTRUCTION INDUSTRY

Contents

Acknowledgements

We are indebted to our friends and colleagues, James Ekins, Jim Padden and Alan Dolman for the helpful suggestions received when writing the first edition of this book and Ian Cooper for his valuable assistance when writing the chapter on Operational Research.

We would also like to express our gratitude for the help and encouragement received from our wives.

Introduction

The book is concerned with the application of management techniques to problems occurring in the industry. It is assumed that readers are familiar with the basic techniques and no excuse is made for omitting much of the background information in some chapters.

The examples used have been simplified essentially to avoid over complication, and adapted to get the important points over to the reader. The techniques used are applied in a simple form avoiding the more sophisticated extensions applied by some firms. The methods of presentation have evolved from our experience with students studying the various topics.

The techniques in this book have been considered separately for convenience. It must be appreciated, however, that if a firm is to work efficiently, it needs an integrated information system which will meet the needs of management. For this to be possible, it is necessary to cross the normal departmental boundaries. Information which is gathered for one department such as production control, may be relevant to other departments such as surveying, cost control, etc. In all firms a certain amount of integration exists but there are few firms indeed which can claim to have a fully integrated system and this includes firms in industry generally.

There seems very little point in departments within a firm collecting information or processing information which is identical to that collected or processed by another department but this happens all too often in practice. In addition to this, much information is collected which is never used or is not presented in an acceptable manner for different levels of management. The system should avoid these deficiencies. It is a fact that many systems in a firm happen more by accident than by design. If an information system is to be designed for a whole business the 'traditional organisation structure' will be replaced by a 'functional structure'. This is necessary because information must flow vertically and horizontally and in a traditional structure information tends to flow vertically, but information flow across the lines of authority is not always present and this results in the problem of duplication of information mentioned earlier.

Space restricts the consideration of a worked example in the design of an integrated system, but an article written by the authors is referred to in the bibliography.

Management techniques are the tools of management, and the techniques considered in this book can and should be used by all firms, irrespective of size. They enable decisions to be taken with greater knowledge of the facts; ignorance of the techniques available to management leads to inefficient operation and may eventually lead to bankruptcy in the competitive years ahead. The use of scientific methods is essential to effective management.

This book is aimed mainly at students studying management as part of a degree, Higher National Diploma and other courses in construction, and for students preparing for the examinations of the Institute of Building. It will also be useful to personnel practising in building who are concerned with production planning and control.

We hope that this book will fill a void which we feel exists in the available text books on management in the construction industry.

Rotherham
April, 1971   R. OXLEY and J. POSKITT

Preface to Second Edition

In preparing the second edition of this book we felt that it was necessary to include additional topics to bring it up to date with recently introduced techniques and to extend some of the existing chapters. The new material includes such topics as Method Statements, Sequence Studies, Multi-Project planning, Time-scale Presentation of C.P.M., The Line of Balance Method of Planning, Incentives and a completely new Chapter on Operational Research.

We hope that the second edition will be as well received as the first edition.

1: Construction Planning and Control

1.1 INTRODUCTION

Planning can be applied in varying degrees of detail, depending on the stage at which it is being carried out. For construction work it is usually divided into pre-tender planning, pre-contract planning, and short term planning on site.

Pre-tender planning is done to allow the estimator to arrive at an estimate of cost based on the proposed methods of working and an estimate of the time required to carry out the work. Programming at the pre-tender stage is in an outline form to consider only the phasing of the main operations, since much information is not at that time available.

Pre-contract planning is carried out when the contract has been won and the project is considered more fully. Planning at this stage includes the overall programme, labour schedule, plant schedule, materials schedules, etc. The overall programme should not break the operations down excessively or it will become unrealistic.

Short term planning on site is done in greater detail and the programmes at this stage are broken down much further.

Control must be carried out to make planning effective, as without control planning loses much of its value. It must be applied continuously to up-date the plans and to enable reconsideration of work ahead in the light of what has already taken place. The example which follows could be programmed using the Critical Path Method. The resulting overall programme, converted to a bar chart, would then have presented much more information, e.g. float times on all activities could be shown. However, it is argued by some that for small projects, bar charts may be drawn direct and are quite satisfactory. Many Contractors use bar charts as they are thought to be less involved. In addition it would not be possible in examination conditions to produce a programme showing movement of resources starting with a network as this would involve drawing a network, analysing it and then scheduling which would involve resource allocation. This could then result in an extension of the time first calculated. Whichever method of programming is used it should be appreciated that when progressing is done the aim is to adhere to the programme.

1.2 PRE-TENDER PLANNING

Pre-tender planning will not be considered in detail but it is felt that the relevance of the pre-tender method statement is too important to omit.

1.2.1 Pre-tender Method Statements

This is one of the most important pre-tender documents because it is used by the estimator as the basis for arriving at the cost of many of the major items of work. It follows that the method statement should include all work of any consequence where alternative methods or choice of plant are possible.

The method statement takes the form of a schedule (see Fig. 1.1.) stating the method to be adopted or plant to be used for such activities as excavation, formwork, concreting, etc., and is compiled at the same time as the pre-tender programme and site layout drawing, having considered in detail each problem and having arrived at the best solution with regard to resources required and cost. Having arrived at the resources required, it is common practice to record these in the schedule for future reference. Together with the pre-tender programme it forms the basis for the overall programme. The schedule is also very useful for informing site staff of the method the estimator has used when arriving at a price.

Time spent in the preparation of a well thought out method statement usually results in a more competitive tender and when considered with a programme it can result in the solution of some potential problems before they arise.

A pre-tender Method Statement for the Amenity Centre and Office Block shown in Fig. 1.1. follows. This method statement is used as the basis for the overall programme, see Fig. 1.8.

Fig. 1.1   PRE-TENDER METHOD STATEMENT

CONTRACT: AMENITY CENTRE   SHEET NO: COMPILED BY:
          & OFFICE BLOCK

NO. OF SHEETS   DATE

Item	Quantity	Remarks	Labour & Plant	Time Req'd Days
Hoarding	45 linear metres	Erected along N.E. boundary of site 2·5 metres high.	2 carpenters	2
Access	— — —	Vehicle access via works entrance Cross over in S.W. footpath Workmen access via 1 metre wide gate in hoarding (position shown on site layout drawing).	2 labourers —	1 —
Accommo-dation	—	All accommodation to be firm's standard sectional huts (position shown on site layout drawing) Toilets provided in form of mobile unit.	2 carpenters	8
Temp. Services	—	Provide electrical, water & W.C. services as shown on site layout drawing.	3 labourers Plumber & mate Electrician & mate	3 } 3 } 2
Excava-tion	105 cubic metres	*Reduced level dig* Front bucket of J.C.B. 4 — easy going.	J.C.B.4 + operator & 3 labourers	1

Fig. 1.1 continued

Item	Quantity	Remarks	Labour & Plant	Time Req'd Days
	140 cubic metres	*Tie beam trenches* Rear back-acter bucket of J.C.B.4. *Internal drain trenches* Rear back-acter bucket of J.C.B.4.	J.C.B.4. & operator & 3 labourers	4
		Disposal of surplus spoil Remove to tip 400 metres distance using lorries — within works.	2 lorries & drivers	5
General Hoisting & Transportation	—	The tower crane will be used for lifting all materials until frame is complete when it will be replaced by a 750 kg platform hoist which will be built adjacent to T.C. before it is removed. N.B. all bricks & blocks lifted onto floor before crane removed. Platform hoist used for lifting materials used in finishings.	Tower crane & driver Hoist	80 80
Concreting	420 cubic metres	*Mixing* The concrete will be mixed in a 300/200 litre electrically powered reversing drum mixer, having a built-in weigh batcher, the aggregate being delivered to the weigh batcher by a hand scraper. All located at south corner of site. Cement is to be stored in a 20 tonne silo located directly over the mixer.	2 labourers 300/200 litre mixer set up	75
	420 cubic metres	*Transportation* Mixed concrete is to be transferred to all levels by means of the tower crane using a 400 litre skip.	Tower crane 500 kg capacity 21 metre max. reach + driver	—
	420 cubic metres	*Placing* The concrete is to be compacted by means of an immersion vibrator.	3 labourers Immersion vibrator	—

Fig. 1.1 continued

Item	Quantity	Remarks	Labour & Plant	Time Req'd Days
Rein-force-ment	37 tonnes	**Work Area** All reinforcing steel to be cut & bent on site in the area set aside at the east corner of the site, adjacent to the existing building. The equipment in this area consists of a scaffold rack for storage of steel when delivered, and a scaffold rack for storage of steel when processed.	3 steel fixers	
	—	**Plant** Cutting bench, bending bench and power bender all on hard standing and covered with tarpaulin on a scaffold frame.	Power bender	62½
	—	**Transport** Steel lifted to the required level using the tower crane.	Tower crane	
Form-work	—	**Working Area** All formwork is to be produced on site in the area set aside at the S.E. end of the building. A joiners' shop to be provided consists of a scaffold tube frame covered with a tarpaulin and having a concrete hardstanding.		
	—	**Plant** Electrically powered circular saw.	Circular saw	
	—	**Transport** Formwork lifted to required level using tower crane.	Tower crane	
	—	**Foundations** Rough boarded formwork to bases and edge of floor slab at ground floor.	4 carpenters	63½
	—	**Columns** 12 mm plywood sheet with 100 mm x 50 mm sawn backing, secured by column cramps (see drawing no.) supported by telescopic props.	8 carpenters	

Fig. 1.1 continued

Item	Quantity	Remarks	Labour & Plant	Time Req'd Days
	—	*Beams* 12 mm plywood sheet with 100 x 50 mm sawn backing to soffite and sides, supported by Telescopic props.		
	—	*Slabs* 12 mm plywood sheet with 100 mm x 50 mm sawn bearers on expanding steel floor centres.		
Mortar-Mixing	—	Mortar for bricklaying will be mixed in a 150/100 litre mixer located on hard standing adjacent to the T.C. or hoist. A bagged cement store will be located alongalongside.	1 labourer	
Scaffolding	—	Scaffolding will be of independent type 1 m wide with staging at 2 m intervals and be constructed of tubular steel. A ramp will be provided at the N.W. end of the building to provide access to all platform levels.	2 scaffolders	

1.3   PRE-CONTRACT PLANNING

1.3.1 Basis of calculations

The information to be used as a basis for calculating productive hours should be obtained from operational records of past contracts and from work study synthetics if these are available. The bill of quantities can be used for the master programme providing it is accurate, but if it is employed many items will have to be collected together under the heading of the programme operations (e.g. the operation 'brickwork' will include such items as 'extra over for facings' and 'bedding cills').

The bill rates should be broken down to show labour and plant content so that the labour and plant content of operations can be easily ascertained.

1.3.2 Breakdown into operations

At this stage the project should be broken down into the major operations only and the temptation to reduce it to detailed operations must be resisted.

1.3.3 Description of project

The project consists of a five storey building with a reinforced concrete frame cast *in situ*. It contains an amenity centre and office accommodation (see Figs. 1.2, 1.3 and 1.4), the amenity centre being confined to the lower three floors whilst the office accommodation is on the upper two floors.

THIRD FLOOR

FOURTH FLOOR

SECOND FLOOR

FIRST FLOOR

GROUND FLOOR

Fig. 1.2 Floor plans—amenity centre and office block

SOUTH EAST ELEVATION

NORTH WEST ELEVATION

NORTH EAST ELEVATION

SOUTH WEST ELEVATION

Fig. 1.3 Elevations—amenity centre and office block

Fig. 1.4 Perspective of amenity centre and office block

1.3.4 Construction of the building

1.3.4.1 Foundations
Reinforced concrete bases and concrete strip foundations.

1.3.4.2 Superstructure
Frame: reinforced concrete columns, beams, floors and roof.
Ground floor: reinforced concrete on D.P.M. on blinding on hardcore filling.
Staircases and lift walls: reinforced concrete cast *in situ*.
External facing: aluminium curtain walling secured to reinforced concrete frame.
Concrete block backing to cill level on inside of curtain walling and 300mm
brick cladding where shown on drawing.
Covering to roofs and canopy: screed laid to falls with asphalt finish.

1.3.4.3 Internal finishes and services
Internal partition walls: 100mm clinker blocks.
Rainwater goods and drainage: in service ducts.
Wall finishes: partial tiling to all toilets, washrooms, locker rooms and drying
rooms. Timber panelling to entrance halls and office accommodation.
Plastering elsewhere with eggshell finish paint.
Ceiling finishes: suspended ceilings to office accommodation. Plastering

elsewhere. All ceilings have eggshell paint finishes.
Floor finishes: Terrazzo to amenity centre and to all staircases and landings.
Wood blocks to office accommodation.
Joinery work: hardwood faced doors, hardwood skirtings to office
accommodation. Softwood internal fittings and general joinery work painted.
W.c. partitions and shower cubicles: specialist metal partitions.
Heating: supplied from an external source within the works.
External column facings: mosaic finish.
Lifts: to office accommodation only.

1.3.4.4 External works
Drainage as shown on drawings.
Boundary walls.
Services.
Tarmac paving on blinding on hardcore.

1.3.5 The site
The site is located 6.00 km from the contractors head office and is
within the boundaries of a chemical works; access is via an existing works
entrance. The ground is level and consists of loose filling to a depth of 1.2 m
Below this is a subsoil of firm gravel.

1.3.6 Contract period
The contract period is 46 weeks, the starting date is February 19th.

1.3.7 Working week
A 40-hour, 5-day week has been used throughout.

1.3.8 Plant
The major plant has been selected at the pre-tender stage as follows:

500 kg static tower crane on concrete base
300/200 litres reversing drum concrete mixer, with loading hopper
    and power scoop
20 tonne cement silo
J.C.B. 4 for all excavation work by machine
500 kg platform hoist for lifting when the tower crane leaves site.

1.3.9 Productive hours
Plant-hours and man-hours (for the productive operatives only) have been
established for all operations (Fig. 1.5).

1.3.10 Sub-contract work
Sub-contractors have been contacted and have agreed periods on site and
notice required (Fig. 1.6).

1.3.11 Procedure for programming
1. Study the drawings before starting the programme.
2. Estimate the time required for major phases of the work, e.g.

Set up site
Sub-structure
R.C. frame
Structural brickwork and curtain walling

Roof covering
Internal trades and finishings
External works
Cleaning up and tidying site.

This will act as a guide when preparing the overall programme and will indicate in good time whether sufficient resources are being used to complete on time.

The planned completion date should be set before the contract period to allow for unforeseen delays.

3. List the operations to be used in the overall programme in approximate order of starting time, grouping the operations shown in the schedules as necessary. This list will contain the operations shown on the calculation sheet (Fig. 1.8) and the overall programme chart (Fig. 1.9).

4. Fill in the total plant-hours or productive man-hours for the programme operations.

5. Decide on the number of operatives or items of plant to be used on the first operation and calculate the time required. Enter this on the programme chart before proceeding to the next operation. Attention must be paid to providing continuity of work for plant and men where possible.

Fig. 1.5   SCHEDULE OF OPERATIONS

Operation	Plant-hours	Man-hours (productive trades)
Erect hoarding		28
Cast tower crane base		13
Set up mixer, including all hard standings		40
Site accommodation		110
Excavate for drains and services		63
Water service and w.c.		20
Electric service		13
Reduce level dig	7 (J.C.B.)	21
Excavate trenches and bases, including internal drains	28 (J.C.B.)	84
Reinforcement in bases, foundations, lift base, and floor slab		98
Rough formwork to bases and edge of floor slab		196
Concrete in bases and trenches	22 (300/200 litre mixer)	
Internal drainage		96
Reinforced concrete frame:		
reinforcement		1402
formwork		4071
concrete	213 (300/200 litre mixer)	

Operation	Plant-hours	Man-hours (productive trades)
Hardcore filling and blinding		152
D.P.M. and concrete ground floor slab	17 (300/200 litre mixer)	
Brickwork: to first floor level		185
first to second floor		264
second to third floor		258
third to fourth floor		264
fourth floor to roof		258
External works:		
reduce level dig	14 (J.C.B.)	
excavate for drains and manholes	10 (J.C.B.)	
excavate for services	7 (J.C.B.)	
excavate for foundations to boundary walls	7 (J.C.B.)	
concrete to foundations; drains and ducts	28 (300/200 litre mixer)	
brickwork to manholes and ducts, and lay drains		21
brickwork to boundary walls		85
backfilling, hardcore and blinding		205
clear huts from site		68
clear up		208
Blockwork backing to curtain walling:		
to first floor level		78
first to second floor		45
second to third floor		45
third to fourth floor		33
fourth floor to roof		80
Internal block partitions:		
to first floor level		174
first to second floor		196
second to third floor		196
third to fourth floor		166
fourth floor to roof		90
Roof screed		72
First fixings:		
carpenter		325
plumber		410
electrician		450
Plastering:		
to walls		590
to ceilings		375
floor screeds		350

Operation	Plant-hours	Man-hours (productive trades)
Wall tiling		847
Second fixings and finishings:		
carpenter		480+150=630
plumber		320
electrician		220
Painting and decorating:		
external		68
internal		840

Fig. 1.6   SCHEDULE OF SUB-CONTRACTORS

Operation	Time required (weeks)	Notice required (weeks)
Erect tower crane	1½	1
Asphalt roof covering	1	8
Curtain walling and glazing	6	12
Heat engineer:		
first fixing	6	16
second fixing	3	
Lift installation	8	20
Suspended ceilings	4	16
Cubicles to w.c.'s and showers	3	11
Timber wall panelling	2	7
Terrazzo floor finish	6	12
Wood block floor finish	2½	18
Mosaic to columns	1	10
Connect and lay services	½	2
Tarmac paving	1	8

1.3.12 Key operations

Key operations are those which determine the time required for the contract, e.g. sub-structure, frame erection, roofed-in stage, etc.

It is essential that key operations are planned realistically. If these are correct, then the programme will be practical.

1.3.13 Sequence studies

1.3.13.1 Introduction

It is necessary to compile sequence studies when closely related operations have to be coordinated, e.g. the construction of the columns, beams and slabs in a reinforced concrete framed building or the finishing trades in a building such as a multi-story office block which would have much repetitive work on the various floors. (The line of balance technique described in section 1.9.21 could also be used.)

1.3.13.2 Achieving continuity
To achieve continuity of work throughout these repetitive cycles, it is necessary to balance the gangs of workmen. It is usual to take the key operation or trade and work out the best cycle for it and then to fit the other operations around this key operation, limiting unproductive time in the other operations as much as possible by balancing gangs. Taking the example of the construction of the reinforced concrete frame of a building, the key operation would be formwork to the frame. Consequently in compiling a sequence study for such a project the first step would be to work out the best cycle of operations for the formwork and then to integrate the steelfixing and concreting operations.

1.3.13.3 Sequence study for the frame of the Amenity Centre
1. Formwork stripping times
The stripping times for formwork stated in the B.O.Q. are as follows:—

Suspended floor and beam soffites	7 days (props left in)
Columns	1 day
Staircases	7 days

To provide continuity of work for carpenters it will be necessary to have sufficient formwork for half the columns on one floor, one complete floor soffite and all the stairs to one floor.

2. Procedure
The procedure for drawing up the sequence study is as follows:—

(a)   Using the quantities involved, calculate the work content for carpenters and steelfixers as follows

Carpenters	4071 hours.
Steelfixers	1402 hours.

(b)   Balance the gangs to avoid non-productive time. It is difficult to provide continuity for the concrete gang

Carpenters
4071 hours — with 8 carpenters = 509 gang hours
Steelfixers
1402 hours — with 3 steelfixers = 467 gang hours

(c)   Using the gang sizes calculated above calculate the work content for each operation (see sequence study chart Fig. 1.7).

(d)   Programme the work providing continuity on the repetitive work cycles on each floor.

(e)   In the case of the 1st floor slab and beams, there is a total of 584 man hours on formwork which for the workforce of 8 carpenters gives a total time requirement of 10.25 days. Steelfixing for this floor slab and beams requires a total of 219 man hours which to give a similar time requirement in days will require 3 steelfixers. It must be borne in mind that the formwork contains a certain amount of time for strip and that the steelfixing contains a certain amount of time for preparation, therefore the actual fixing times will still be approximately equal and therefore the balance is maintained. It is not always possible to maintain complete balance between these trades throughout the project if the same relationship between trades is maintained. An example of this occurs on staircases where the relative time requirements are 3.25 days to 1.75 days. Here the steelfixers could be fabricating steel for columns and beams required later in the project.

3. General notes
   (a) As operatives become experienced in the repetitive operations the output would increase so that on the upper floors, working time would be reduced. This has not been allowed for in the programme.
   (b) Scaffolding has not been shown on the programme but this would be programmed in sequence with the frame.
   (c) The reinforcement cages for the columns and some of the beams are to be prefabricated. The tower crane will be used to lift the fabricated columns and beams into position and to move formwork from floor to floor.
   (d) Weeks only have been used on Fig. 1.7 for clarity in publication but days would also be included in practice.

1.3.14 Flexibility

To allow for wet time, absenteeism, etc., a 35-hour week has been used in the calculations and flexibility has also been provided by not overlapping operations the maximum time possible. Overtime working could be used to gain time if necessary.

1.3.15 General notes on overall programme (Fig. 1.9)

1. For operations in first six weeks, see notes with Six-weekly programme (p. 32).

2. Setting out

Setting out will be carried out during first week.

3. Frames

Carpenters on formwork will determine the time required for erection of the frame. The steelfixing gang will be balanced with the carpenter gang so that the time taken for steelfixing is approximately the same as that for formwork.

The staircases and lift walls are cast as work proceeds.

The floors are used to store bricks at each level before the next floor formwork is erected.

4. Brickwork and blockwork

Start of brickwork and blockwork is delayed until erection of frame is well advanced. This will also give continuity of work for bricklayers.

5. First and second fixing

Gang sizes are selected to provide continuity of work from first to second fixing. In the case of carpentry, it may be the firm's policy to have carpenters specializing in first and second fixing.

6. Plastering and tiling

Sufficient plasterers and tilers are used to overlap first fixing but to allow continuity of work for trades moving onto second fixing.

7. Suspended ceilings

Follow plastering to upper floors.

8. W.c. cubicles

Fixed to plastered or tiled walls and floor.

9. Wall panelling
Starts after commencement of wood blocks.
10. Floor finishes
Wood block, starts well after screeds to allow them to dry out.
11. Mosaic finishes to columns
Delayed to end of project to avoid damage.
12. Painting and decorating
External painting done first to allow scaffold to be stripped.

1.3.16 Crane

The crane will be released from site as soon as heavy lifting is complete. Whilst on site, the programme should provide maximum use, and to do this the crane must be programmed daily if it is to be used effectively. The gang sizes can be greatly influenced by use of a crane on a site.

1.3.17 Hoist

The hoist is introduced to take over vertical movement when the crane leaves the site.

1.3.18 Labour and plant

Labour and plant have been assumed available when required, but in practice availability could be affected by the requirements of other projects. In this case the programme would take account of availability and this could delay the start of operations.

1.3.19 Plant for general movement of materials

This has not been included on this programme or in the plant schedule (Fig. 1.11).

1.3.20 Notice required by sub-contractors

The calling-up dates indicated on the programme are shown 1 week earlier; this is to give sub-contractors the full notice required.

1.3.21 Labour schedule (Fig. 1.10)

This is drawn up on the basis of the overall programme by adding up the labour required each week. It shows the approximate number of each trade required and the total number of the contractors' men on site.

A minimum of two labourers is kept on site for unloading and general sweeping and cleaning up, and two extra labourers have generally been provided, the only exceptions being in week numbers 3 and 4 and week numbers 35 to 40, when it is felt that extra labourers would not be economical.

1.3.22 Plant schedule (Fig. 1.11)

This is also drawn up from the overall programme and shows the major items of plant required each week. Dismantling times have not been included.

1.3.23 Other schedules

Material requirement schedules, sub-contractors schedules, information requirement schedules, and component requirement schedules can be drawn up, based on the overall programme.

1.3.24 Financial control

A financial graph can be prepared from the overall programme by applying costs to the operations. This will provide a means of financial control.

	OPERATION		PLANT HRS.	PROD MAN HRS.	LAB REQD	TIME REQD (DAYS)
GD. TO 1ST. FLOOR	COLUMNS	REINF		78	2/3 S	4
		FWK		240	8 C	4.25
		CONC	14			2
	1ST FLOOR	FWK		584	8C	10.25
		REINF		219	3S	10.25
		CONC	31			4.5
	STAIRS	FWK		178	8C	3.25
		REINF		36	3S	1.75
		CONC	9			1.25
1ST. TO 2ND. FLOOR	COLUMNS	REINF		75	3S	3.5
		FWK		202.5	8 C	3.5
		CONC	9.5			1.25
	2ND FLOOR	FWK		543	8 C	9.75
		REINF		194	3S	9.25
		CONC	27			4
	STAIRS	FWK		178	8C	3.25
		REINF		36	3S	1.75
		CONC	9			1.25
2ND. TO 3RD. FLOOR	COLUMNS	REINF		75	3S	3.5
		FWK		202.5	8C	3.5
		CONC	9.5			1.25
	3RD FLOOR	FWK		543	8C	9.75
		REINF		194	3S	9.25
		CONC	27			4
	STAIRS	FWK		178	8C	3.25
		REINF		36	3S	1.75
		CONC	9			1.25
3RD.TO 4TH. FLOOR	COLUMNS	REINF		75	3S	3.5
		FWK		202.5	8C	3.5
		CONC	9.5			1.25
	4TH FLOOR	FWK		543	8C	9.75
		REINF		194	3S	9.25
		CONC	27			4
	STAIRS	FWK		98	8C	1.75
		REINF		18	3S	1
		CONC	5			0.75
4TH. FLOOR TO ROOF	COLUMNS	REINF		48	3S	2.25
		FWK		125	8C	2.25
		CONC	9.5			1.25
	ROOF	FWK		322	8C	5.75
		REINF		120	3S	5.75
		CONC	14			2
	STAIRS	FWK		32	8C	0.5
		REINF		6	3S	0.25
		CONC	3			0.5

Fig. 1.7 Sequence study for amenity centre R.C. frame.

Fig. 1.8   CALCULATIONS SHEET

Operation no.	*Operation*	*Plant-hours*
1	**Set up site**	
	Erect crane	
2	**Excavation**	
	Reduce level dig	7 ⎫
		⎬ 35
	Excavate trenches, bases and internal drains	28 ⎭
3	**Formwork in frames bases,and slab**	
	Rough formwork to bases and edge of floor slab	
	In R.C. frame	
4	**Reinforcement in frames and bases**	
	Reinforcement in bases, foundations and floor slab	
	In R.C. frame	
5	**Internal drainage**	
6	**Concrete in frames, bases and trenches**	
	Concrete in bases and trenches	22 ⎫
		⎬ 235
	In R.C. frame	213 ⎭
7	**Ground floor construction**	
	Hardcore fill and blinding	
	D.P.M. and concrete ground floor slab	17
8	**Brickwork cladding**	
	to 1st floor level	
	1st to 2nd floor	
	2nd to 3rd floor	
	3rd to 4th floor	
	4th floor to roof	
9	**Curtain walling and glazing**	

Productive man-hours	Labour and plant	Time required (weeks)	Remarks
	2 carpenters 2/3 general labourers Plumber and mate Electrician and mate	2	Setting out for dig also undertaken in this period. G.P.O. to connect up for phones
	Crane erection gang (sub-contractor)	1½	
21 ⎫ ⎬ 105 84 ⎭	J.C.B. 4 and operator 3 labourers 2 lorries and drivers	1	Reduce level dig to loose fill can be moved with J.C.B. front bucket. Labourers dig widenings and bottom-up
196 ⎫ ⎬ 4267 4071 ⎭	4 carpenters for two weeks, then 8 carpenters Tower crane	16	
98 ⎫ ⎬ 1500 1402 ⎭	3 steel fixers Tower crane	15	
96	2 plumbers and mates	1¼	
	300/200 litre mixer set up Tower crane 5 labourers, then 4 labourers	7	Work spread over 15½ weeks. Floors loaded with bricks
152	7 labourers ¾ ⎫ 300/200 litre mixer set up ⎬ 1¼ 5 labourers ½ ⎭		Work spread over a period of 2 weeks and integrated with bases and trenches
185 ⎫ 264 ⎪ 258 ⎬ 1229 264 ⎪ 258 ⎭	6 bricklayers 4 labourers 150/100 litre mixer	6	1 labourer on 150/100 litre mixer
	Sub-contractor	6	12 weeks notice required

Fig. 1.8 continued

Operation no.	Operation	Plant-hours
10	**Roof screed**	
11	**Asphalt**	
12	**Internal blockwork to all levels** to 1st floor level 1st to 2nd floor 2nd to 3rd floor 3rd to 4th floor 4th floor to roof	
13	**Internal block partitions** to 1st floor level 1st to 2nd floor 2nd to 3rd floor 3rd to 4th floor 4th floor to roof	
14	**Carpentry** 1st fixing 2nd fixing	
15	**Plumbing** 1st fixing 2nd fixing	
16	**Electrical work** 1st fixing 2nd fixing	
17	**Heating** 1st fixing 2nd fixing	
18	**Plastering and floor screeds** Plastering to walls Plastering to ceilings Floor screeds	
19	**Wall tiling**	
20	**Lift installation**	

Productive man-hours	Labour and plant	Time required (weeks)	Remarks
72	2 plasterers 2 labourers 150/100 litre mixer	1	
	Sub-contractor	1	8 weeks notice required
78 45 45 } 281 33 80	2 bricklayers 1 labourer 150/100 litre mixer	4	
174 196 196 } 822 166 90	4 bricklayers 2 labourers 150/100 litre mixer	6	
325 630	2 carpenters 2 carpenters	5 9	
410 320	2 plumbers and mates 2 plumbers and mates	6 4½	
450 220	1 electrician and mate 2 electricians and mates 1 electrician and mate	4 4½ 6	4 weeks total time for conduits laid on formwork before concreting by 1 electrician and mate
	Sub-contractor Sub-contractor	6 3	16 weeks notice required
590 375 } 1255 300	6 plasterers 4 labourers 150/100 litre mixer	6	
847	6 tilers 3 labourers	4	
	Sub-contractor	8	20 weeks notice required

Fig. 1.8 continued

Operation no.	Operation	Plant-hours
21	**Suspended ceilings**	
22	**Cubicles to w.c.'s and showers**	
23	**Wall panelling**	
24	**Floor finishes** Terrazzo	
25	Wood blocks	
26	**Painting and decorating** External Internal	
27	**Mosaic finishes**	

EXTERNAL WORKS

28	**Excavation** Reduce level dig Excavate for drains and manholes Excavate for services Excavate for foundations to boundary walls	14 10 7 7 } 38
29	**Concrete foundations, beds to drains, ducts**	28
30	**Brickwork to manholes and boundary walls** Brickwork to manholes and ducts, lay drains Brickwork to boundary walls	
31	**Connect and lay services**	
32	**Backfill hardcore and blinding**	
33	**Tarmac paving**	
34	**Clear huts from site (includes hoardings)**	
35	**Generally clean and tidy-up**	

Productive man-hours	Labour and plant	Time required (weeks)	Remarks
	Sub-contractor	4	16 weeks notice required
	Sub-contractor	3	11 weeks notice required
	Sub-contractor	2	7 weeks notice required
	Sub-contractor	5	Toilet areas executed first to allow sanitary fittings to be fixed; 12 weeks notice required
	Sub-contractor	2½	18 weeks notice required
65	2 painters	1	
840	6 painters	4	
	Sub-contractor	1	10 weeks notice required
42 ⎫ 30 ⎪ 114 21 ⎬ 21 ⎭	J.C.B. 4 and operator 3 labourers 2 lorries and drivers	1	Labourers dig branches and generally bottom-up
84	150/100 litre mixer 3 labourers	¾	
21 ⎫ ⎬ 106 85 ⎭	2 bricklayers 1 labourer 150/100 litre mixer	1½	
	Sub-contractor	½	2 weeks notice required; laid by personnel from chemical works
205	3 labourers, vibrating roller for compacting	2	
	Sub-contractor	1	8 weeks notice required
68	2 carpenters	1	
208	3 labourers	2	

OVERALL BUILDING PROGRAMME FOR AMENITY CENTRE & OFFICE BLOCK   **CONTRACT NO.**

OP. NO.	OPERATION	MONTH	FEB	MARCH	APRIL	MAY	JUNE	JULY	AUGUST	SEPTEMBER	OCTOBER	NOVEMBER	DECEMBER	JANUARY	FEB
		WEEK COMM.	19 26	4 11 18 25	1 8 15 22 29	6 13 20 27	3 10 17 24	1 8 15 22 29	5 12 19 26	2 9 16 23 30	7 14 21 28	4 11 18 25	2 9 16 23 30	6 13 20 27	3 10
		WEEK NO.	1 2	3 4 5 6	7 8 9 10 11	12 13 14 15	16 17 18 19	20 21 22 23	24 25 26 27	28 29 30 31 32	33 34 35 36 37	38 39 40 41	42 43 44 45	46 47 48 49	50 51 52
1	SET UP SITE														
2	EXCAVATION			2/3 GENERAL LABS. J.C.B. 4 & 6 OPERATOR, 2 LORRIES & DRIVERS. 2 CARPS. PLUMB. & MATE. ELECT. & MATE.				CRANE ERECTION GANG IN 3RD & 4TH WEEKS. SUB-CONTRACT						
3	FORMWORK IN FRAMES BASES & SLAB							3 GENERAL LABS. INTEGRATED WITH OP. NO. 1							
4	REINF. IN FRAMES & BASES							4/8 CARPS. TOWER CRANE & GANG.							
5	INTERNAL DRAINAGE				2 PLUMB. & MATE.		2/3 STEELFIXERS. TOWER CRANE & GANG.								
6	CONC. IN FRAMES BASES & TRENCHES				(1 FROM OP. NO.1 FROM OP. NO.1)			THEN 4 LABS (CRANE GANG) 500 kg TOWER CRANE							
7	GROUND FLOOR CONSTRUCTION				100/100 MIXER & 7 LABS INTEGRATED WITH OP. NO. 6			100/100 MIXER & 5 LABS. – LORRIES SPREAD H.C. AS DELIVERED							
8	BRICKWORK CLADDING							6 BRICKLAYERS 4 LABS. 150/100 MIXER, HOIST AFTER HOLIDAY							
9	CURTAIN WALLING								SUB-CONTRACT						
10	ROOF SCREED							2 PLAST. 2 LABS. 150/100 MIXER, HOIST							
11	ASPHALT			SCAFFOLD REC'D				SUB-CONTRACT.							
12	INTERNAL BLOCKWORK TO CILL LEVEL							2 BRICKLAYERS 1 LAB. 150/100 MIXER, HOIST, AFTER HOLIDAY.							
13	INTERNAL BLOCK PARTITIONS					1ST FIX – CONDUITS OR FORMWORK		2 BRICKLAYERS 4 BLAYERS, 2 LABS, 150/100 MIXER, HOIST (2 OFF FROM OP. NO.12)							
14	CARPENTRY								2 CARPS, HOIST						
15	PLUMBING							1ST FIX 2ND FIX	2 PLUMBERS & MATES, HOIST						
16	ELECTRICAL WORK							1ST FIX 2ND FIX	2/1 ELECTICIANS & MATES						
17	HEATING							1ST FIX 2ND FIX	SUB-CONTRACT						
18	PLASTERING & FLOOR SCREEDS								6 PLASTERERS 4 LABS, 150/100 MIXER & HOIST						
19	WALL TILING								6 TILERS 3 LABS						
20	LIFT INSTALLATION								SUB-CONTRACT						
21	SUSPENDED CEILINGS								SUB-CONTRACT						
22	CUBICLES TO W.C.s & SHOWERS								SUB-CONTRACT						
23	WALL PANELLING								SUB-CONTRACT						
24	FLOOR FINISHES – TERRAZZO						ROOF WATER-TIGHT WALLS WATERTIGHT		SUB-CONTRACT.						
25	WOOD BLOCK								SUB-CONTRACT.						
26	PAINTING & DECORATING							2 PAINTERS	INTERNAL 6 PAINTERS						
27	MOSAIC FINISHES								SUB-CONTRACT						
	EXTERNAL WORKS.														
28	EXCAVATION							START STRIPPING SCAFFOLD		J.C.B. 6 OP. 2 LORRIES & DRIVERS. 3 LABS					
29	CONC. FDNS & BEDS TO DRAINS ETC.									3 LABS 150/100 MIXER					
30	BRICKWORK TO M.H.s & BOUNDARY WALLS									2 BRICKLAYERS 1 LAB 150/100 MIXER					
31	CONNECT & LAY SERVICES.									SUB-CONTRACT					
32	BACKFILL, HARDCORE, & BLINDING.														
33	TARMAC PAVING									SLABS. VIBRATING ROLLER					
34	CLEAR HUTS FROM SITE									SUB-CONTRACT					
35	GENERALLY CLEAN & TIDY UP.									3 LABS					
					BANK HOLIDAY ANNUAL HOLIDAY	BANK HOLIDAY ANNUAL HOLIDAY	ANNUAL HOLIDAY		BANK HOLIDAY			ANNUAL HOLIDAY HOLIDAY	PERIOD		
	CALLING UP DATES SHOWN THUS △														

Fig. 1.9

LABOUR SCHEDULE FOR AMENITY CENTRE & OFFICE BLOCK.

TYPE OF LABOUR.
CARPENTERS
GENERAL LABOURERS
PLUMBER & MATE
ELECTRICIAN & MATE
STEELFIXERS
SCAFFOLDERS
BRICKLAYERS & LABOURERS
PLASTERERS & LABOURERS
TILERS & LABOURERS
PAINTERS & DECORATORS
TOTALS

PLANT SCHEDULE

TYPE OF PLANT
500 Kg. MIXER SET-UP (INCLUDING SILO)
J.C.B. 4 & OPERATOR
4 m³ LORRIES & DRIVERS.
500 Kg. TOWER CRANE & OPERATOR
POWER BENDER
CIRCULAR SAW
500 Kg. PLATFORM HOIST
150/100 MIXER
VIBRATING ROLLER
400 LITRE DUMPER

Fig. 1.10 (top) and Fig. 1.11 (lower)

1.4   DETAILED METHOD STATEMENT

Method statements can be compiled along with stage or short term programmes recording in detail the methods selected for completing a particular stage of the work on a project. The information in this type of method statement would be primarily for the benefit of the site staff, but it could be fed back to the estimator for his future reference.

An example of a more detailed method statement for set-up site and the construction of the sub-structure of the Amenity Centre and Office Block is shown in Fig. 1.12.

Fig. 1.12   DETAILED METHOD STATEMENT
                          SET UP SITE AND FOUNDATIONS

CONTRACT   SHEET NO.   COMPILED BY
NO. OF SHEETS   DATE:

Item	Quantity	Remarks	Output Per Hour	Labour & Plant	Time Req'd (Days)
		SET UP SITE			
Erect hoarding	45 linear metres	To be erected along the N.E. boundary of the site only to a height of 2·5 m. Paving to be removed within hoarding. Construction of hoarding to be as shown in temporary works drawing No. . A gate 1 m wide to be provided at N.W. end for personnel and visitor access.	1.4	2 carpenters	2
Tower crane base	6½ cubic metres	*Excavation* Excavate for tower crane base by hand, dimension to be as drawing No.	0.5	2 labourers	
	4½ cubic metres	*Concreting* Base to be concreted using truck mixed concrete discharged straight from truck mixer standing on works service road. Mild steel angles cast into base to receive bottom section of tower crane mast.	2.25	2 labourers	1
	—	*Disposal of spoil* Spoil to be left alongside excavations and removed later with reduced level dig.	—	2 lorries + drivers J.C.B.4. + operator	

Fig. 1.12 continued

Items	Quantity	Remarks	Output Per Hour	Labour & Plant	Time Req'd (Days)
Hard-standings	—	Hard-standing to be con-creted using readymixed concrete delivered via works service road. Con-crete for mixer hard standing delivered via truck shute, conc. for other hard-standings delivered by wheel-barrow. Excavated spoil to be placed alongside hard-standing area for removal with reduced level dig.	—	3 labourers	} 2
Set-up Mixer	—	Mixer and silo delivered to site on low loader off loaded and located as shown on site layout. Mobile crane hired from client for off-loading and erection of silo. Walls between aggregate to be constructed from railway sleepers and rolled steel I section stakes.	—		
Site Accom-modation	—	To be located in areas shown on site layout plan. Sectional timber construction based on 1 m module. All provid-ed with insulation, heating and lighting. All units to be supported clear of the ground on railway sleepers.	2	2 carpenters	2
All services	21 cubic metres	*Excavation* Excavation for all temporary services to be carried out by hand, any surplus material being deposited along-side the excavation for removal at same time as reduced level dig.	3	3 labourers	3

Fig. 1.12 continued

Items	Quantity	Remarks	Output Per Hour	Labour & Plant	Time Req'd (Days)
	—	*Water service & W.C.* Service provided to areas as indicated on site layout plan — water pipes to be P.V.C. and placed 300 mm below surface. W.C.'s provided in mobile amenity unit located over and connected to existing m.h. — special cover (see drawing No. 365) used to make gas-tight.	—	Plumber & mate	3
	—	*Electrical service* Electrical service provided in form of 415 V three phase supply to all electrically operated plant and 220 V single phase to accommodation for heating and lighting. Location of supply to be as shown on site layout drawing. Supply to messroom from dist. board in clerk of works office to be by overhead cable. All other electrical service routes to be underground in trench 300 mm deep — protected by pitch fibre pipe.	—	Electrician & mate	2
Erect Tower Crane	—	Erection of crane to be carried out by fitters from plant department.		Crane erection gang	7
Foundations	105 cubic metres	*Reduced Level dig* Reduced level dig will be carried out over area of building by J.C.B.4. using front bucket, as going is light — all spoil will be loaded into lorries and taken to tip within works, total haul distance approx. 400 metres.	15	J.C.B.4. & operator 3 labourers	1

Fig. 1.12 continued

Items	Quantity	Remarks	Output Per Hour	Labour & Plant	Time Req'd (Days)
	140 cubic metres	*Excavate Trenches and bases* Tie beam trenches excavated using rear back-acter bucket of J.C.B.4. surplus spoil loaded into lorries and removed to tip. Bottoming up and trimming at isolated foundation to be completed by hand.	5	J.C.B.4. & operator 3 labourers	4
	See Bar Schedule for quantities	*Reinforcement* Reinforcement for foundations prepared in cutting and bending area and carried to foundation by hand.	–	2 steel fixers	7
	See Formwork Schedule for quantities	*Rough formwork* Formwork to foundation and edge of slab to consist of sawn boards and bearers, prefabricated in site shop and transported to fixing position by hand x square metres being used, stripped and refixed as required.	–	4 carpenters	7
Internal Drainage	See Drainage Schedule for quantities	All internal drainage is in cast iron – trench excavation carried out by J.C.B. along with excavations for tie beams – transported from stores area by hand, surrounded with granular fill when fixed.	–	Plumber & mate	7
Concrete in bases & trenches	66 cubic metres	Concrete produced in site batching plant and delivered to bases etc. by dumper.	3	300/200 litre mixer + operator & 7 labourers	3

Fig. 1.12 continued

Items	Quantity	Remarks	Output Per Hour	Labour & Plant	Time Req'd (Days)
Hardcore fill	100 cubic metres	Limestone hardcore delivered to site in 10 tonne lorries and tipped into place. Sleepered area provided to bridge tie beams thus allowing lorries to pass into area within tie beams. Limestone spread from back of lorries as it is tipped, final levelling being carried out by hand.	10	7 labourers	} 3½
Blinding	14 cubic metres	Concrete for blinding delivered from batching plant by dumper tipped into position and spread and levelled by hand.	1	300/200 litre mixer + operator & 7 labourers	
Concrete Ground Floor	51 cubic metres	*Mixing* Mixed in site batching plant. *Transportation* From batching plant to forms by dumper *Reinforcement* Mesh reinforcement placed in centre of slab. *Placing* Compaction carried out using vibrating tamping beam. *Construction* Hit and miss construction – stop ends placed as shown on sketch below (Fig.1.13).	3	300/200 litre mixer + operator & 7 labourers	} 2½

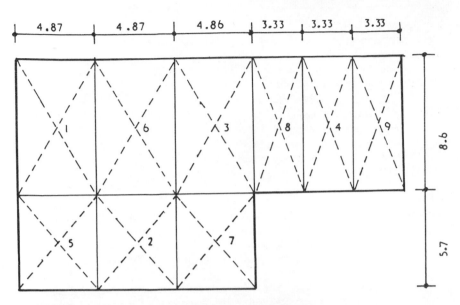

Fig. 1.13 Location of construction
joints in ground floor slab

1.5 SHORT TERM PLANNING

1.5.1 Period of time covered

Short term programming can cover a period of four to six weeks or may
cover a stage of the work. The programme considered here will cover a period
of six weeks. A new programme is drawn up every fourth week, thus giving
a two week overlap and allowing a review to be made of the work outstanding
from the previous month; any work behind schedule can then be included
in the current programme.

1.5.2 Purpose

The purpose of short term programming is to ensure that work proceeds in
accordance with the overall programme. The overall programme is thus converted
into a working schedule and must be updated regularly, whilst a review is
simultaneously made of all requirements by checking schedules.

1.5.3 Time of preparation

The most suitable time for preparation of the programme is after the
monthly planning meeting, which is in turn best held immediately after the
architect's monthly meeting. In this way account can be taken of any alterations
or variations which are necessary.

1.5.4 Originator of programme

The programme should be originated by the site manager, if necessary with
the assistance of a projects manager and a planning engineer. The co-operation
of foremen and sub-contractors should be obtained by getting them involved
and keeping them informed. They would be present at the monthly site meeting
to give their point of view and agree the general plan of action.

1.5.5 Degree of detail necessary

The operations in the overall programme will have to be broken down to show each element of work to be completed, e.g. the reinforced concrete frame — which was one operation on the overall programme — will be broken down into the elements of work on each floor:

> Reinforcement in columns
> Formwork to columns
> Concrete to columns
> Formwork to beams and slab
> Reinforcement to beams and slab
> Concrete to beams and slab.

1.5.6 Basis of calculations

The basis of calculations at this stage must be the operational records from previous contracts. There should be an integrated cost control/bonusing/planning and surveying system which will provide the basic information for all departments. The expected outputs for bonus operations should be used in calculating the time required.

The bill of quantities cannot be used in calculating time periods for these detailed operations as the outputs used to price the bill are averages and do not take into account the conditions of work in individual operations.

1.5.7 The six-weekly programme (Fig. 1.14)

The period covered is the first six weeks of the contract.

1.5.8 Procedure for programming

1. Break down the operations in the overall programme into more detailed operations as shown in the six-weekly programme (Fig. 1.14), taking care to include all operations. Any preparatory work necessary must be included.

2. Fill in plant-hours and/or productive man-hours for the detailed operations.

3. Decide on the number of operatives and plant necessary and calculate the time required, plotting each operation on the programme as it is considered. The aim is to provide continuity of work and to meet the targets set in the overall programme.

1.5.9 Labour requirements

These can now be obtained from the six-weekly programme by adding up the labour required each day (Fig. 1.14). Extra labourers necessary for off-loading, etc. are not included here.

1.5.10 Plant requirements

These can also be obtained from the six-weekly programme.

1.5.11 Other requirements

Requirements for details, nominations, materials, etc. can also be obtained from this programme and a check made with the schedules originated at the overall programming stage to ensure that all requirements will be available when required.

1.5.12 General notes on six-weekly programme

1.5.12.1 Water for works and drinking

During the first week this can be obtained from a tap on the works near to the site. After the first week, it will be laid on.

1.5.12.2 Tower crane base and all hardstandings
Ready mixed concrete will be used for pouring the tower crane base and
hardstandings.

1.5.12.3 Cutting, bending and fixing steel
The steelfixers' time includes cutting and bending. Fixing of steel proceeds
after rough formwork in bases and slab, with steel in bases and trenches being
fixed first to allow concreting to proceed.

1.5.12.4 Rough formwork in bases and slab
Formwork to bases has priority so that concreting can proceed.

1.5.12.5 Plant-hours
It has been assumed that the plant fixes the speed of operations for excavating
and concreting and that sufficient labour is used to attend the plant.

1.5.12.6 Formwork
The two-day period between 'rough formwork in bases and slab' and 'formwork
to columns' can be used for adjusting the form panels, fixing stops, etc.

1.5.12.7 Use of crane
When the crane is not used on concreting operations it will be employed for
moving formwork, placing steel, distributing materials, etc., in accordance
with the daily programme (Fig. 1.17)

1.5.12.8 Stripping times for columns
It is assumed that the stripping time for columns is 24 hours. The hours given
includes erect, strip, clean, and re-oil.

1.5.12.9 Concreting to first floor, including canopies
Concreting to first floor includes the canopies and is placed in two sections.
The first section is placed when half the formwork and steel is fixed, thus
allowing half the formwork to be stripped earlier and re-used on the next floor.

1.5.13 Sub-contract programmes
If required these can be extracted from the detail programmes for handing to
sub-contractors and show how the sub-contractor's work ties in with the main
contractor's work.

1.6 WEEKLY PLANNING

1.6.1 Purpose
The purpose of weekly programming is to ensure that the six-weekly
programme is effectively carried out. It is used as a basis for operational
instructions, for communicating the plan to the trades foremen, gangers,
and operatives, and to help in co-ordinating the requirements of
different sites.

1.6.2 Time for preparation
It should be prepared after a weekly site meeting at which gangers and foremen
are present, together with others who will be involved in the following week's
work.

| CONTRACT-AMENITY CENTRE & OFFICE BLOCK. CONTRACT NO. | | SIX WEEKLY PROGRAMME | | PERIOD -FEB 19TH TO MAR 29TH |

OP. NO	OPERATION	PLANT HRS or PROD M.HRS	LABOUR &/OR PLANT	TIME REQ'D DAYS	FEBRUARY / MARCH
	SET UP SITE				
1	ERECT HOARDING	28	2 CARPS	2	
2	TOWER CRANE BASE	13	2 LABS.	1	
3	SET UP MIXER & ALL H'D'ST'GS	40	3 LABS.	2	
4	SITE ACCOMMODATION	110	2 CARPS	8	
5	EXCAVATE FOR ALL SERVICES	63	3 LABS.	3	
6	WATER SERVICE & W.C.	20	PL.& M'TE	3	
7	ELECTRIC SERVICE	13	EL.& M'TE	2	
8	ERECT TOWER CRANE	ERECT'N GANG		7	← CONC. MATURING →
	FOUNDATIONS				
9	REDUCE LEVEL DIG	7	JCB.& OP 2 LO RR	1	
10	EXC. TRENCHES & BASES	28	3 LABS	4	
11	REINF. IN FOUNDS. ETC.	98	2 ST'F'X'S	7	
12	R'O. FWK. IN BASES & SLAB	196	4 CARPS	7	
13	INTERNAL DRAINAGE	96	2PL.&MTE	7	
14	CONC. IN BASES & TRENCH'S	22	MXR& 7L	3	INTEGRATE WITH OP 15
15	H.C. FILL & BLINDING	152	7 LABS	3½	INTEGRATE WITH OP 14
16	D.P.M. & G.F. CONC.	17	MXR+7L	2½	
	COLS TO U/S. F.F. BEAMS				
17	REINF.	78	2/3 ST'F'X'S	4	
18	FORMWORK	240	8 CARPS	4½	
19	CONCRETE	14	MIXER & CRANE+4	2	
	1ST. FLOOR INC. CANOPY				
20	FORMWORK	584	8 CARPS	10½	
21	REINF.	253	3 ST'F'X'S	12	
22	CONCRETE	31	MIXER & CRANE+4	4½	
23	CONDUITS (IN FL. SLAB)	27	1EL.&MTE	4	
24	SCAFFOLDING				

REMARKS	LABOUR REQUIREMENTS	
	CARPS.	2 — 2 — 4 — 4 — 8 — 8 — 8
	GENERAL LABS.	2 — 3 — 3 — 3 — 7 — 7 — 4 — 4
	PLUMBER & MATE	1+1 — 2+2 — 2+2
	ELECT. & MATE	1+1 — 1+1
	STEELFIXERS	2 — 2 — 3 — 3 — 3
	CR. ERECT'N GANG	SUB-CONTRACT
	SCAFFOLDERS	2 — 2 — 2 — 2

REMARKS	PLANT REQUIREMENTS	
	300/200 MIXER	
	J.C.B. 4 & OPER'T'R	
	4 m³ LORR. & DR. (2)	
	500 Kg. TOWER CR. & OP	
	POWER BENDER	
	CIRCULAR SAW	
	400 litre DUMPER	

Fig. 1.14

1.6.3 Originator of programme

The programme should be originated by the site manager with the assistance of a planning engineer if necessary. The work should be discussed with foremen and·sub-contractors when they are affected.

1.6.4 Degree of detail necessary

Considerable detail is necessary here, so the programme will show days and sometimes even hours.

1.6.5 Basis of calculations

As with the six-weekly programme, the basis of calculations should be operational records. Since weekly programming is tied up very much with the bonus system, the expected outputs for bonus purposes are probably the best basis and can be evolved from work study synthetics from the work study department. Information from bonus feed-back may be used to up-date this programme.

1.6.6 Procedure for programming (Fig. 1.15)

1. List all the operations to be carried out during the next week in the order of starting time.

2. Insert the quantity of men and plant required and (taking into account the work content) enter the time required on the programme, providing continuity of work wherever possible.

1.6.7 Labour requirements

The labour requirements are entered below the weekly programme and this will assist in co-ordinating the requirements of different sites.

1.6.8 Plant and other requirements

Plant and other requirements should now be finally checked to ensure that they will be available when required.

1.6.9 General notes on weekly programme

1.6.9.1 Erect reinforcement to columns
This operation is the completion of the fixing of the column cages which were started the previous week. The crane would assist as necessary in lifting cages into position.

1.6.9.2 Erect formwork to columns
The carpenters follow the steelfixers to complete the column formwork started the previous week.

1.6.9.3 Pour concrete to columns
The crane gang follows the carpenters erecting column forms. Crane and gang leave this operation for short periods to lift column cages.

1.6.9.4 Strip formwork to columns
Stripping of the columns which were poured first can start as soon as the last column forms are erected. This will leave plenty of time for concrete to harden.

1.6.9.5 Prepare reinforcement to beams and floors
The steelfixers cut, bend, and fabricate steel for floors after reinforcement to columns has been erected.

OP NO	OPERATION	LABOUR & PLANT	DAY	MONDAY	TUESDAY	WEDN'DAY	THURSDAY	FRIDAY	SATURDAY	SUNDAY
	CONTRACT: AMENITY CENTRE & OFFICE BLOCK CONTRACT NO.		WEEKLY PROGRAMME					WEEK NO. 5 WEEK COMM. MAR 18.		
			DATE	MAR. 18 TH	19TH	20TH	21ST	22ND	23RD	24 TH
1	ERECT REINF. TO COLS.	3 STEELFIXERS								
2	ERECT FWK. TO COLS	8 CARPS								
3	POUR CONC TO COLS	4 LABS. 300/200 MIXER								
4	STRIP FWK TO COLS.	8 CARPS								
5	PREPARE REINF TO BEAMS & FLOOR	3 STEELFIXERS								
6	ERECT FORMWORK TO BEAMS & FLOOR SLAB	8 CARPS.								
7	FIX REINFORCEMENT TO BEAMS & FLOOR	3 STEELFIXERS								
8	ERECT SCAFFOLDING	2 SCAFFOLDERS								
	LABOUR									
	CARPENTERS	PLANNED		8	8	8	8	8		
		ACTUAL								
	STEELFIXERS	PLANNED		3	3	3	3	3		
		ACTUAL								
	LABOURERS	PLANNED		4	4	4	4	4		
		ACTUAL								
	SCAFFOLDERS	PLANNED		2	2	2	2	2		
		ACTUAL								

Fig. 1.15

1.6.9.6 Erect formwork to beams and floor slab
The carpenters move onto this after stripping columns; crane will assist by lifting formwork panels into position. The floor will be cast in two equal sections to allow earlier release of some of the formwork.

1.6.9.7 Fix reinforcement to beams and floors
The steelfixers follow carpenters on floors, fixing steel. The crane is used to lift beam cages as necessary in accordance with the programme for the crane.

1.6.9.8 Erect scaffolding
Scaffolders erect scaffold to the outside of the building as work proceeds.

1.6.9.9 Crane and operator
The tower crane and operator have been omitted from the weekly programme as a separate programme for the crane will be drawn up each day.

1.6.10 Communicating the programme
There are many methods of communicating the programme to trades foremen, gangers and operatives, including pictorial diagrams, sequence studies, simplified bar charts, and work lists; the method chosen will depend upon the type of work being undertaken.

The operation and location must be clearly defined. On small projects the work of several gangs can be shown on one sheet or diagram, but on large projects a separate sheet or diagram will be necessary for each trade or gang.

The weekly site meetings are of great assistance in communicating a programme.

1.6.11 Sequence studies
The sequence study should be adhered to when drawing up programmes on site.

1.6.12 Simplified bar chart (Fig. 1.16)
A simplified bar chart will be used to communicate the weekly programme for the amenity centre and office block.

CONTRACT: AMENITY CENTRE & OFFICE BLOCK.						WEEK NO. 5 WEEK COMM MAR 18	
CONTRACT NO.	MONDAY	TUESDAY	WEDNESDAY	THURSDAY	FRIDAY	SATURDAY	SUNDAY
CARPENTERS (8)	ERECT FORMWORK TO COLUMNS.	STRIP FORMWORK TO COLUMNS		ERECT FORWORK TO BEAMS & FLOOR SLAB			
STEELFIXERS (3)	ERECT REINF TO REST OF COLS	PREPARE REINFORCEMENT TO FLOORS		TO BEAMS &	FIX REINF TO BEAMS & FLOORS		
LABOURERS (4)	POUR CONCRETE TO ALL COLUMNS		WORK AS DAILY PROGRAMME FOR CRANE				
SCAFFOLDERS (2)	ERECT SCAFFOLD OUTSIDE BUILDING TO ALLOW POURING OF CONCRETE						

Fig. 1.16

The work of each trade or gang is set down on one bar line, and the chart is read in conjunction with an outline layout drawing which can be used for each floor. The plan is referenced so that the extent of work to be carried out in the week can be clearly shown, foremen and gangers being given a copy of both the chart and the diagrams.

1.7 DAILY PROGRAMMING FOR THE TOWER CRANE

1.7.1 Planning in detail
The work of the tower crane must be planned in detail if it is to be used effectively.

If the tower crane is to be employed on one operation all day it may be necessary for the crane and gang to work overtime the previous day, preloading the scaffold with materials for other trades.

1.7.2 The work of the crane gang
When the crane gang (four labourers) are not all working on operations with the crane, any surplus men can be preloading pallets, preparing loads, etc.

1.7.3 The daily programme (Fig. 1.17)
The programme is for Thursday, 21st March. It is assumed that progress is as programmed on the weekly site plan.

CONTRACT AMENITY CENTRE & OFFICE BLOCK	DAILY PROGRAMME FOR TOWER CRANE	DAY : THURSDAY DATE : 21 ST MARCH.
OPERATION	GANG	7.00 8.00 9.00 10.00 11.00 12.00 1.00 2.00 3.00 4.00 5.00 6.00 7.00
STRIP COLUMN FORMWORK	CARPENTERS	
ERECT FORMWORK TO FIRST FLOOR	CARPENTERS	
LOAD UP GROUND FLOOR	4 LABOURERS	

Fig. 1.17

1.7.4 General notes

   1. Strip column formwork

The crane is working with carpenters, mainly moving formwork to be cleaned and re-oiled for its next application.

   2. Erect formwork to first floor

Again working with the carpenters, the crane is engaged in lifting beam and column formwork into position ready for fixing. Carpenters are then left to fix the formwork lifted whilst the crane moves to the next operation, subsequently returning to assist the carpenters.

   3. Load up ground floor

Bricks are loaded on the ground floor while it is still accessible to the crane. The loading areas will be pre-determined to avoid obstructing formwork supports.

   The crane is programmed to service specific gangs or operatives throughout the day, thus avoiding a situation where everyone requires the crane at the same time. On the day shown on the programme, the operations listed are the only ones which require the use of the crane: it will carry out any lifting required by the gangs involved during their time allocation.

1.8. CONTROL OF PROGRESS ON PROJECTS

1.8.1 Introduction

   Control is complementary to planning, therefore progressing is complementary to programming.

   Control involves comparing, at regular intervals, the actual achievement with the plans and then taking any necessary corrective action to bring things back on schedule.

1.8.2 Use of meetings

   The monthly and weekly meetings mentioned earlier in this chapter are invaluable in helping to control progress. Work which is not proceeding as planned will receive particular attention, and explanations will be required where sufficient progress is not being achieved. The action necessary for correcting under-production will be considered and the best solution will then be incorporated into the programme for the next period.

1.8.3 Effective control

   If a programme is to be really effective as a control document, it must represent time and quantity of work done. This is particularly important on programmes where operations take several weeks or months to perform, since time alone will not clearly indicate whether work in progress is behind plan or not.

1.8.4 Methods of recording progress

   Progress can be recorded on pictorial diagrams by colouring plans and elevations when certain sections of work are completed. This method is often used on housing projects and gives a quick visual impression of overall progress, but other ways are also available depending on the type of building and the amount of detail necessary.

   In this instance progress will be shown on the planning charts, which clearly indicates what is happening and where corrective action will have to be taken.

1.8.5 Progressing the weekly programme (Fig. 1.18)

   The example given shows the progress of operations at 5 p.m. on Wednesday, 20th March.

CONTRACT : AMENITY CENTRE & OFFICE BLOCK CONTRACT NO.		WEEKLY PROGRAMME				WEEK NO 5 WEEK COMM. MAR.18				
OP. NO	OPERATION	LABOUR & PLANT	DAY DATE	MONDAY MAR.18TH	TUESDAY 19TH	WEDN'DAY 20TH	THURSDAY 21ST	FRIDAY 22ND	SATURDAY 23RD	SUNDAY 24TH
1	ERECT REINF. TO COLS.	3 STEELFIXERS								
2	ERECT. FWK TO COLS.	8 CARPS.								
3	POUR CONC. TO COLS	4 LABS 300/100 MIXER								
4	STRIP FWK TO COLS.	8 CARPS								
5	PREPARE REINF. TO BEAMS & FLOOR.	3 STEELFIXERS.								
6	ERECT FORMWORK TO BEAMS & FLOOR SLAB	8 CARPS.								
7	FIX REINFORCEMENT TO BEAMS & FLOORS.	3 STEELFIXERS								
8	ERECT SCAFFOLDING	2 SCAFFOLDERS								

LABOUR						CURSOR		KEY PROGRAMME WK.COMPLTD TIME LINE	
CARPENTERS	PLANNED	8	8	8	8	8			
	ACTUAL	8	8	8					
STEELFIXERS	PLANNED	3	3	3	3	3			
	ACTUAL	2	2	2					
LABOURERS	PLANNED	4	4	4	4	4			
	ACTUAL	4	4	4					
SCAFFOLDERS	PLANNED	2	2	2	2	2			
	ACTUAL	2	2	2					

Fig. 1.18

1.8.6 Use of cursor

A cursor is used to show the position reached on the time scale.

1.8.7 The programme line

The programme line is used to represent time and quantity of work.

1.8.8 The time line

The time line is used to indicate the time at which an operation starts and finishes.

1.8.9 Analysis of progress

1. Erect reinforcement to columns

This operation is now complete but took half a day longer than planned because only two steelfixers were available instead of the three planned.

2. Erect formwork to columns

This operation is now complete, but also took half a day longer than planned due to previous operation finishing late.

3. Pour concrete to columns

This operation is now complete. It took slightly longer than planned because of the formwork finishing late.

4. Strip formwork to columns

This operation started late because the carpenters were delayed on operation number 2. The production of the carpenters is slightly higher than expected as they have done 25 per cent more work than planned.

5. Prepare reinforcement to beams and floor

This operation started late because of the delay in operation number 1. There are still only two steelfixers available and work is falling behind plan.

1.8.10 Taking corrective action

It is obvious from the information given that unless another steelfixer is obtained the work on the frame will be delayed and the carpenters will not be able to work to capacity. The site manager would therefore contact the projects manager to get a steelfixer as soon as possible. It may in fact be necessary to get two steelfixers for a short period to catch up on work behind plan.

By progressing the plan, corrective action can be taken before serious delays occur.

1.8.11 Up-dating the six-weekly programme

The six-weekly programme can be up-dated at weekly intervals from the weekly site plan and a review made of the operations in the coming week to see if they are affected.

1.8.12 Up-dating the overall programme

This can also be up-dated weekly from the weekly site plan, and will show the overall position of progress to date.

1.8.13 Keeping head office informed

This can be done by means of a shuttle programme which is up-dated by the site manager each week.

Alternatively a progress report can be sent to head office giving the percentage completion of all operations, with reasons for delay and any other relevant information.

1.9   PROGRAMMING FOR REPETITIVE CONSTRUCTION

1.9.1 Introduction

When a project or part of a project consists of operations which are repeated many times, the programme must (as in bar charts for non-repetitive projects) attempt to provide continuity of work for the operatives and plant. The most obvious example of this situation is in housing projects, where the units have more or less the same work content whether they are two-storey houses, maisonettes, or flats. Another example is formwork, reinforcement, and concrete in multi-storey reinforced concrete buildings.

1.9.2 Balancing of gangs

In repetitive construction, the work of the different trades or gangs must take approximately the same period of time to complete, otherwise unnecessary non-productive time results.

1.9.3 Example of the use of balanced gangs

A number of identical buildings are to be built, the operations involved, together with their work content, being as follows:

Excavation	4 man-days
Concrete foundations	16 man-days
Brickwork	32 man-days
Roof construction	12 man-days
Roof finishings	8 man-days

If, in manning the operations, the need to provide the same period of time for each trade is ignored, the result may be as follows:

Operation	Work content (days)	Number of men to be used	Period of time taken (days)
Excavation	4	2	2
Concrete foundations	16	2	8
Brickwork	32	6	5⅓
Roof construction	12	2	6
Roof finishings	8	4	2

This would result in the programme shown in Fig. 1.19, which shows that some trades have considerable non-productive time as follows:

Bricklayers are non-productive for 2⅔ days in 8 days
Carpenters are non-productive for 2 days in 8 days
Roofers are non-productive for 6 days in 8 days

Fig. 1.19

Whilst it may be possible to employ these men on other sites, the gangs could equally well have been balanced to give continuity of work for all the operatives:

Operation	Work content (days)	Number of men to be used	Period of time taken (days)
Excavation	4	2	2
Concrete foundations	16	4	4
Brickwork	32	8	4
Roof construction	12	3	4
Roof finishings	8	2	4

This would result in the programme shown in Fig. 1.20.

Excavation is planned on a faster cycle but since it is the first operation this does not result in non-productive time and the men engaged on it can leave the site on completion. Obviously some flexibility is essential and this can be provided by delaying the start of operations on housing which follows (as shown in the example).

Fig. 1.20

1.9.4 Description of a housing project

An example will now be considered on a housing project consisting of an estate of 30 pairs of semi-detached houses. The site layout and working drawings are shown in Figs 1.21 and 1.22.

1.9.5 Construction of houses

1.9.5.1 Foundations
Concrete strip foundations, brick cavity walls up to damp-proof course level.

1.9.5.2 Ground floor
Hardcore filling, blinded with ashes. Damp-proof membrane, site concrete (ground floor slab).

1.9.5.3 Superstructure
External walls: cavity constructed with brick outer leaf and clinker block inner leaf.
Internal garage wall: brickwork.
All other internal walls: clinker blocks.
First floor tongued and grooved, boarding on timber joists.
External features: cedar boarding on battens on building paper on clinker blocks.

1.9.5.4 Roof construction
Traditional construction with T.D.A. roof trusses.
Eaves finish: fascia board and soffit board.
Verges: barge boards.

1.9.5.5 Roof coverings and rainwater goods
Main roof: concrete slates on battens on felt.
Garage projections: nuralite on boarding on firrings on joists;
All rainwater goods: cast iron.

1.9.5.6 Internal finishes and services
Bathroom and kitchen walls: partial tiling.
Other walls: two coats of plaster.
Ceilings: plaster board and skim.
Heating: gas fired boiler
Fireplace: stonework built on site.
Staircase: traditional softwood.

Ground floor finishes: wood blocks on screed to dining room and hall;
P.V.C. tiles on screed to kitchen.
Joinery work including fittings: softwood.
Soil pipes: cast iron.
Painting: knot prime stop and three coats of oil paint on all softwood, three
coats of emulsion paint on all walls.

1.9.5.7 External works
Paving to drive and around houses.
Rustic fencing.

1.9.6 The site
The site is located four miles from the contractor's office, and is easily
accessible. The ground is level and the sub-soil is sandy clay.

When the site was acquired by the contractor, the roadworks, together with
drains and mains, had been completed.

1.9.7 Working week
A 40-hour, 5-day week has been used throughout, and working days only
have been shown on the programme.

1.9.8 Productive hours
Plant-hours and man-hours (for productive operations only) have been
established for all operations (Fig. 1.23).

Fig. 1.21

Fig. 1.22 Housing at South Cove, Tenbury
General arrangement

Fig. 1.23 HOUSING AT SOUTH COVE, TENBURY

Table 1: Schedule of operations

Operation no.	Operation	*Productive trade/skill or plant*	*Hours per pair for productive trade/skill or plant*
	Sub-structure up to D.P.C.		
1	Preliminaries	2 carpenters, plumber and mate, electrician and mate, 2 general labourers	35 hours total for site
2	Excavate over site, remove spoil 30 m and deposit	J.C.B.	Average 14
	Excavate trenches for foundations and level, and ram bottoms	NOTE: easy going assumed, therefore use front bucket of J.C.B.	
3	Concrete in foundation trenches	300/200 litre mixer	Average 7
4	Brickwork to D.P.C. (includes filling cavity)	Bricklayer	Average 80
5	Hardcore and backfill (includes levelling and ramming service entries, blinding, d.p.m.	Labourer	Average 66
6	Over site concrete (solid ground floor)	300/200 litre mixer	6
	Super-structure		
7	Brickwork: 1st lift (includes door and window frames)	Bricklayer	166
8	Scaffolding	Scaffolder	105
9	Brickwork: 2nd lift (includes lintels and beams	Bricklayer	166
10	First floor joists and roof to garage	Carpenter	40
11	Brickwork: 3rd lift (includes window frames)	Bricklayer	110
12	Brickwork: 4th lift (includes lintels, chimney, making good putlog holes and around pipes)	Bricklayer	64
13	Roof carcass (includes barge boards, fascia, soffit, etc.)	Carpenter	71
14	Brickwork top-out (includes beam filling and raking cutting etc.)	Bricklayer	48
15	Roof coverings (includes battens and felt)	Roofer (sub-contractor)	40

Fig. 1.23 continued

Operation no.	Operation	Productive trade/skill or plant	Hours per pair for productive trade/skill or plant
16	Flashings, gutters, downpipes and soilstack (includes garage projection with nuralite roof covering)	Plumber	29
17	Glazing	Plumber	25
	External works		
18	Excavate for drains and services (includes assisting with backfill)	J.C.B. and labourers	Average 27
19	Build manholes, lay drains and test; build stopcock boxes	Bricklayer	Average 48
20	Service mains		
21	Backfilling	Labourer	Average 30
22	Paving, fences, generally clean and tidy site	Labourer	Average 103
	Internal trades and finishings		
23	Plumber: 1st fixing (pipe carcass etc; includes gas pipes)	Plumber	28
24	Heating engineer: 1st fixing (pipe carcass etc.)	Heating engineer (sub-contractor)	60
25	Electrician: 1st fixing (conduit carcassing)	Electrician	21
26	Fix stone fireplace built *in situ*	Bricklayer	14
27	Floor boards and staircase	Carpenter	40
28	Plasterboard to ceilings (includes 12 mm plasterboard soffit to garage)	Plasterer	45
29	First floor partitions (includes external cills)	Bricklayer	95
30	Carpenter: 1st fixing (linings, grounds etc.)	Carpenter	36
31	Electrical drops and switchboxes	Electrician	7
32	Plastering to walls and ceilings, screed floors	Plasterer	272
33	Wood block floors	Floor layer (sub-contractor)	140
34	P.V.C. tile floor	Floorlayer (sub-contractor)	20
35	Plumber: 2nd fixing (sanitary fittings etc.)	Plumber	14

Fig. 1.23 continued

Operation no.	Operation	Productive trade/skill or plant	Hours per pair for productive trade/skill or plant
36	Tiling to bathroom and kitchen	Plasterer	53
37	Carpenter: 2nd fixing (includes doors, architraves, skirtings, cupboards etc.)	Carpenter	112
38	Electrician: 2nd fixing (wiring and boards)	Electrician	16
39	Heating boiler, valves to radiators, and radiator brackets	Heating engineer (sub-contractor)	48
40	Make good generally	Plasterer	7
41	Painting (internal and external)	Painter	260
42	Final fixings: electric-light holders	Electrician	3½
43	plumbing (chains and seats)	Plumber	3½
44	radiators	Heating engineer	3
45	furniture	Carpenter	30
46	Clean floors and scrub out	Cleaners	28
47	Final inspection	Site manager and client	

1.9.9 Sub-contract work

The sub-contractors on this project have worked for the contractor previously and have agreed periods on site for each operation. They have also agreed to work to the programme produced providing sufficient notice is given.

1.9.10 Procedure for programming

1. Study the drawings and other information available before starting the programme.

2. List operations in the order of starting time in each stage of the project.

3. Calculate total number of productive man-hours for each trade or skill in each stage of the work (see first two columns in Fig. 1.24). This is done because operations within each stage are dependent on each other, as they were in the previous example (section 1.9.3).

4. Using optimum gang sizes, or man-power available if this is limited, calculate the period of time required by each trade or skill.

The trade or skill requiring the longest period of time will determine the pace for the whole project, and in this example it is bricklaying, which requires 8 days. In housing the key trade is usually bricklaying, plastering or carpentry.

5. By adjusting gang sizes, balance the gangs so that the period of time required by each is approximately equal to that required by the key trade (wherever possible) and fill-in labourers required to assist tradesmen (Fig. 1.24).

A two-day cycle has been used for the work in sub-structure up to D.P.C., but this will not affect the other stages as this is the first stage to be carried out. A four-day cycle has been used for external works, as this is to some extent independent of the construction of the houses.

6. Programme the work, providing as much continuity of work as possible and providing flexibility (Figs. 1.25-1.29).

Fig. 1.24 HOUSING AT SOUTH COVE, TENBURY

Table 2: Productive hours in each stage after balancing gangs

Operation	Hours per pair for productive trade/skill or plant
Sub-structure up to D.P.C.	
Excavate oversite and trenches for foundations	Average 14
Concrete in foundation trenches	7 ⎫
	⎬ 13
Oversite concrete	6 ⎭
Brickwork to D.P.C.	80
Hardcore fill and backfill	66
Superstructure	
Bricklayer	
1st lift	166 ⎫
2nd lift	166 ⎪
3rd lift	110 ⎬ 554
4th lift	64 ⎪
top-out	48 ⎭
Scaffolding	105
Carpentry	
First floor joists and roof to garage	40 ⎫
	⎬ 111
Roof carcass	71 ⎭
Roof coverings	40
Plumbing, glazing, flashings, soil stack and rainwater goods	29 ⎫
	⎬ 54
Glazing	25 ⎭
External works	
Excavate for drains and services	Average 27
Build manholes, lay drains and test, build stopcock boxes	Average 48

Labour and plant required	*Time required (days)*	*Notes*
J.C.B. and driver, 1 labourer	2	
300/200 litre mixer, 2 labourers at mixer, 400 litre dumper and driver, 1 labourer placing concrete	2	
6 bricklayers 3 labourers 200/150 litre mixer	2	
5 labourers	2	
Gang A: 3 bricklayers Gang B: 3 bricklayers Gang C: 2 bricklayers Gang D: 2 bricklayers Gang D: 2 bricklayers All gangs assisted by 5 labourers and 200/150 litre mixer	55 hours = 8 55 hours = 8 55 hours = 8 32 hours ⎱ 24 hours ⎰ 56 hours = 8	Total of 10 bricklayers take 8 days, each gang working on lifts shown and taking 8 days
2 scaffolders	8	
Gang A; 2 carpenters	8	
Sub-contractor, 2 roofers, and 1 labourer	3	
Gang A: 1 plumber and mate	8	
J.C.B. and driver, 5 labourers, 150/100 litre mixer	4	
2 bricklayers, 1 labourer, 150/100 litre mixer	3½	

Fig. 1.24 continued

Operation	Hours per pair for productive trade/skill or plant

Service mains
    Backfilling   Average 30 ⎫
    Paving, fences and generally clean   ⎬ 133
    and tidy site   Average 103 ⎭

Internal trades and finishings
  Plumber
    1st fixing   28 ⎫
    2nd fixing   14 ⎬ 45½
    Final fixing   3½ ⎭

  Heating engineer
    1st fixing   21 ⎫
    Heating boiler, valves to radiators,
    radiator brackets   48 ⎬ 72
    Radiators   3 ⎭

  Electrician
    1st fixing   21 ⎫
    Drops and boxes   7 ⎬ 45
    2nd fixing   14
    Light holders etc.   3½ ⎭

  Bricklayer
    Stone fireplace   14 ⎫
              ⎬ 109
    1st floor partitions   95 ⎭

  Carpenter and joiner
    Floor boards and staircase   40 ⎫
    1st fixing   36 ⎬ 218
    2nd fixing   112
    Door furniture   30 ⎭

  Plasterer
    Plasterboard   45 ⎫
    Plaster walls, screed floors   272 ⎬ 377
    Making-good   7
    Tiling to bathroom and kitchen   53 ⎭

  Flooring
    Wood block   140 ⎫
              ⎬ 160
    P.V.C. tile   20 ⎭

  Painting   260

  Cleaners   28

Labour and plant required	Time required (days)	Notes
5 labourers	4	
Gang B: 1 plumber and mate	6½	
Sub-contractor, 2 heating engineers, 4 labourers	5	
1 electrician and mate	6½	
Gang E: 2 bricklayers, 1 labourer	8	
4 joiners	8	Gang C on 2nd fixing, gang B on remainder
7 plasterers, 4 labourers 150/100 litre mixer	8	Gang B on plastering and screed floors, gang A on remainder
Sub-contractor, 3 floorlayers 1 labourer	8	
5 painters	7½	
2 cleaners	2	

PROGRAMME FOR HOUSING AT SOUTH COVE.

SUB.—STRUCTURE UP TO D

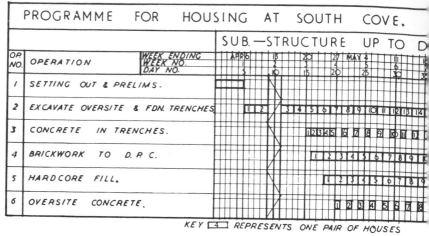

OP. NO.	OPERATION	WEEK ENDING / WEEK NO. / DAY NO.
1	SETTING OUT & PRELIMS.	
2	EXCAVATE OVERSITE & FDN. TRENCHES	
3	CONCRETE IN TRENCHES.	
4	BRICKWORK TO D.P.C.	
5	HARDCORE FILL.	
6	OVERSITE CONCRETE.	

KEY ☐ REPRESENTS ONE PAIR OF HOUSES

TYPE OF LABOUR	LABOUR SCHEDULE FOR
GENERAL LABOURERS. & PLANT DR.VS	
BRICKLAYERS	
B'LAYERS LABS.	

Fig. 1.25

PROGRAMME FOR HOUSING AT SOUTH COVE,

SUPER—STRUCTURE

OP. NO.	OPERATION	WEEK ENDING / WEEK NO. / DAY NO.
7	BRICKWORK 1ST LIFT. — GANG A (3)	
8	SCAFFOLDING	
9	BRICKWORK 2ND LIFT — GANG B (3)	
10	FIRST FLOOR JOISTS ETC. — GANG A (2)	
11	BRICKWORK 3RD. LIFT — GANG C (2)	
12	BRICKWORK 4TH LIFT — GANG D (2)	
13	ROOF CARCASS — GANG A (2)	
14	BRICKWORK TOP OUT — GANG D (2)	
15	ROOF COVERINGS	
16	FLASHINGS, GUTTERS. DOWNPIPES & SOIL STACK — GANG A (1+1)	
17	GLAZING. — GANG A (1+1)	

TYPE OF LABOUR.	LABOUR SCHEDULE FOR
GENERAL LABOURERS	
BRICKLAYERS	
BRICKLAYERS LABOURERS.	
SCAFFOLDERS.	
CARPENTER	
PLUMBER & MATE	
ROOFER & LABOURER	

Fig. 1.26

NBURY

C

JB-STRUCTURE UP TO D.P.C.

NBURY.

SUPER-STRUCTURE

EXTERNAL   WORKS

OP. NO.	OPERATION	WEEK ENDING / WEEK NO. / DAY NO.								
18	EXCAVATE FOR DRAINS & SERVICES									
19	BUILD M.H.'S LAY DRAINS ETC.									
20	SERVICE MAINS									
21	BACKFILL									
22	PAVING FENCING & TIDY SITE									

TYPE OF LABOUR.	LABOUR SCHEDULE FOR EX
GENERAL LABOURERS & PLANT DRIVERS.	
BRICKLAYERS	
BRICKLAYERS LABOURERS	
SERVICES	

Fig. 1.27

1.9.11 Flexibility

Flexibility can be provided in many ways. Typical examples are:

1. Reducing the overlap on operations or stages of work, so that delays in operations or stages do not immediately affect those following.

2. Using a longer time cycle for the various stages of work, e.g. a two-day cycle on foundations, a four-day cycle on superstructure and a six-day cycle on finishings.

3. Leaving any site works out of the main programme so that they can be done in periods of over production.

In this programme flexibility has been provided by reducing the overlap between stages (and in some cases between operations) and by using a faster time cycle for sub-structure up to D.P.C.

Overtime working will be introduced where trades fall behind the programme, but if this is persistent the gang sizes may prove to need adjustment.

1.9.12 Wet time and absenteeism

To allow for wet time, absenteeism, etc., a 35-hour, 5-day week has been used in the programme calculations.

1.9.13 Intermittent working

Where there is insufficient work to keep a particular gang occupied the full 8 days on one unit, the work of that gang must be done intermittently and where this occurs it is better to provide a reasonable amount of work for each visit to the site; it may, of course, be possible to provide continuous work on two sites working simultaneously. The agreement of sub-contractors should always be obtained to the programme.

1.9.14 Rate of selling

The rate at which the houses are expected to sell is another factor which must be considered when deciding on the time cycle of production. In this case it is assumed that selling the houses presents no problem.

NAL WORKS

1.9.15 Alterations to standard design

Alterations to the design must be limited as excessive alterations will disrupt the programme. This would have the effect of unbalancing the gangs, causing non-productive time, and therefore an increase in cost.

1.9.16 Non-productive time

Where the work content for a particular gang does not take the full time cycle for the stage of work, some non-productive time is probable.

1.9.17 Accuracy of outputs

A detailed programme like the one shown is only possible where the operation times can be established with a reasonable degree of accuracy. Feedback information should be available from previous work and can be used as a basis, whilst work study synthetic times can be extremely useful in building up accurate output rates.

1.9.18 Availability of gangs

Continuity of work from site to site will be aimed for in the overall programming policy, so the availability of gangs from previous projects would have an influence on the programme in practice. It has been assumed that gangs can start on site when required.

1.9.19 General notes on programme (Figs 1.25–1.29)

1.9.19.1 Setting out

Setting out will be undertaken by the site manager as the work proceeds, and the first blocks will be set out during the first week of the programme.

1.9.19.2 Sub-structure up to D.P.C.

The reasons for using a two-day cycle on sub-structure up to D.P.C. are:
1. To employ the J.C.B. to the fullest extent whilst it is on site, and

Fig. 1.28

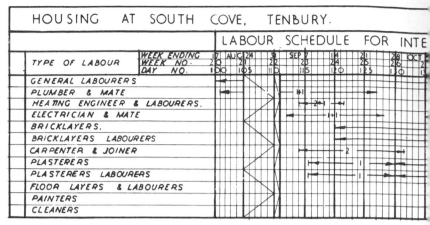

TYPE OF LABOUR	WEEK ENDING	17 AUG 24	31	SEP 7	14	21	28 OCT
	WEEK NO.	20 21	22	23	24	25	26 2
	DAY NO.	100 105	110	115	120	125	130 13
GENERAL LABOURERS							
PLUMBER & MATE							
HEATING ENGINEER & LABOURERS.							
ELECTRICIAN & MATE							
BRICKLAYERS.							
BRICKLAYERS LABOURERS							
CARPENTER & JOINER							
PLASTERERS							
PLASTERERS LABOURERS							
FLOOR LAYERS & LABOURERS							
PAINTERS							
CLEANERS							

HOUSING   AT   SOUTH   COVE,   TENBURY.

LABOUR SCHEDULE   FOR INTE

Fig. 1.29

2. Because work up to D.P.C. can vary considerably from block to block and it is consequently necessary to keep this faily well in front of the work above D.P.C. (by using a two-day cycle it will move quickly in front of work above D.P.C. and does not therefore interfere with the production). When the oversite concrete is completed to block no. 26, 5 labourers plus the dumper driver stay on site until the completion of hardcore, after which 2 labourers leave the site so that 3 labourers plus the dumper driver remain.

1.9.19.3 Plant drivers
The mixer driver has not been included as a plant driver in the labour schedule.

1.9.19.4 Brickwork
The gangs of bricklayers have been kept on the same operation — and sized to take the full time cycle — on each block, thus making the bonusing of these men easy to operate. This method will be acceptable to the men providing the work content of each stage is reasonably accurately assessed. If it is felt necessary to keep one gang on one block, the bricklayers will be split into gangs of 5.

The brickwork in the ground floor partitions is to be built with first and second lift brickwork, whilst the work in first floor partitions is to be undertaken after the floor boards have been laid. Making good the brickwork after the scaffold is removed and around pipes is to be carried out by bricklayers gang D.

1.9.19.5 Roof carcass
The fixing of barge boards, fascia boards etc. has been delayed until after the bricklayers top out.

1.9.19.6 Gangs working on more than one stage
It is of course possible for a gang in a particular trade to work on two stages of the work if there is insufficient work within one stage to take up the full time cycle, e.g. if one gang of bricklayers did not have the full 8 days work on the carcass they could also work on partitions on the first floor. However, it is sometimes desirable to separate trades into specialisms e.g. carpenter and joiner specialising in either carcass, first fixing, second fixing or finishings. This depends mainly on the size of the firm and the amount of work carried out.

It is assumed here that one gang of carpenters will work on floor boards, staircase, first fixing and finishings.

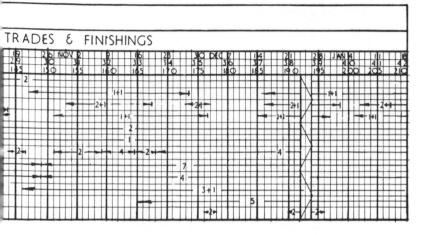

TRADES & FINISHINGS

1.9.19.7 External works

External works will be carried out in sections to give continuity of work to the gangs involved during the periods on site. The excavator is followed by 5 labourers carrying out any necessary hand digging to branches, and concreting beds to drains and manholes.

1.9.19.8 Internal trades and finishings

The bricklayers working on partitions also build the stone fireplace if they are capable of work of this type; if they are not, the fireplaces would have to be built intermittently by a specialist in stonework.

1.9.19.9 Labour schedules

Labour requirements are shown for each stage of the work and indicate clearly the degree of work continuity achieved in the programme. More continuity will be available for some trades when the first block is nearing completion.

The labour required for preliminaries has not been included in the labour schedules, apart from general labourers.

Two general labourers have been kept on throughout the project period for tidying the site, off-loading and general assistance. If necessary they will assist with any hand digging of trenches and bottoming up of trenches in the sub-structure.

1.9.20 Requirements for plant, materials, etc.

When the programme has been drawn up it is possible to determine requirement schedules for plant, materials, drawings etc., in addition to the labour requirements shown on the charts.

1.9.21 The line of balance method

1.9.21.1 The advantages of the method

One of the main advantages of this method is that it gives a better indication of the dependence of one activity on another. It is very useful when progressing because it is immediately obvious when corrective action needs to be taken.

1.9.21.2 Examples of the line of balance method

The housing project described in paragraph 1.9.4. will be used to illustrate the Line of Balance Method. In this example it is assumed that the project is a Local

Authority development and the handover rate required is four houses per week. External works and drainage have been omitted for clarity.

1.9.21.3 Avoiding cyclic queues

In repetitive construction it is best to arrange the schedule so that operatives work on one activity or a group of successive activities on each unit. This is not always possible for small firms producing houses at the rate shown in the previous programme, i.e. one house every four days. However, the problem can sometimes be overcome by grouping activities together even when slow rates of production are essential.

When operatives work on separate activities, on the same unit, cyclic queues are formed as shown in Fig. 6.40. This results in an increase in non-productive, time.

In the following example cyclic queues have been avoided.

The principle of balancing gangs still applies in the Line of Balance Method in order to provide continuity of work for the gangs involved.

1.9.21.4 Procedure for programming

1. Group the activities shown in Fig. 1.23 wherever practicable. The hours for sub-structure and superstructure are for a pair of houses, as it is not practicable to consider these activities separately for each house. The hours for internal work and finishings are for single houses (see Fig. 1.30 columns 1 & 2).

2. Draw a network diagram showing the sequence of activities. A straight line network has been used to avoid overlap of the activities (see Fig. 1.31.

3. Using minimum number of operatives, or items of plant, calculate the gang or plant hours for each grouped activity (see Fig. 1.30 columns 3 & 4).

4. The handover rate required would result in one house being handed over every $35 \div 4 = 8\frac{3}{4}$ hours, once the first house is completed. By considering the gang and plant hours in Fig. 1.30 it can be seen that this handover rate will result in a considerable amount of non-productive time. The hours show that production of one house every 7 hours will be far more economical as most activities can be balanced based on this. The result will be a handover rate of 1 house per day or 5 houses per week and this has been adopted in the schedule. The sub-structure has been scheduled at a rate of 1 per day or 10 houses per week as at this stage the work content in each activity can vary considerably.

5. Some activities will be out of balance and adjustments to gang sizes or amount of plant may be necessary to bring them into balance, e.g. the time for first floor joists was originally based on a gang of two carpenters, this has been adjusted to two carpenters and one labourer.

6. Calculate the number of gangs or items of plant required to give the hand-over rate required, i.e. $\dfrac{\text{gang or plant hours}}{7\ \text{(hours per unit)}}$

This will balance the gangs as nearly as possible (see Fig. 1.30, column 5).

7. Using the number of gangs or items of plant, calculated in 6 above, calculate the number of completions per week for each activity, i.e. $\dfrac{35g}{G}$, where calculations are based on a 35 hour week, g = number of gangs or items of plant and G = gang or plant hours (see Fig. 1.30, column 6).

8. Decide on the number of gangs or items of plant to be used on each unit. (P) (see Fig. 1.30, column 7).

9. Calculate the time required to complete each unit, i.e. $\dfrac{G}{7P}$, (see Fig. 1.30, column 8).

10. Calculate the overall time from the start of a particular activity on the first unit to the start of that activity on the last unit, i.e. $\frac{5(N-1)}{R}$, where N = total number of units (see Fig. 1.30, column 9).

11. Plot the line of balance schedule based on the above information, allowing stage and/or activity buffers as necessary to allow flexibility in the schedule.

NOTE: When the activity preceding the one being considered has a smaller overall time, the buffer will be plotted at unit 1. If the overall time is greater, the buffer will be plotted at the last unit (see Fig. 1.32).

1.9.21.5 *Overlapping activities*
In practice, some activities can proceed concurrently as shown in Fig. 1.33.

The line of Balance Programme for this network is shown in Fig. 1.34.

It can be seen that the programme is much more difficult to read and is more difficult to progress. A good method when using this type of network is to draw concurrent activities on separate transparent sheets and lay them over the line of balance programme or produce separate programmes for these activities.

1.9.21.6 *An alternative method of allowing buffers between activities*
Instead of showing buffers separately for each activity, the activity times can be increased to allow for the buffers. This results in a much simpler schedule, see Fig. 1.35. However, the period allowed for the buffer is not obvious using this method and this may be considered a disadvantage.

1.9.21.7 *Factors governing handover rate*
The handover rate for Local Authority work is usually determined by the Local Authority. If the required handover rate does not fit in with the natural rhythm of the activities, to be performed, either it must be adjusted as in this example, or non-productive time will result, thereby increasing direct costs.

For private development, the forecast rate of selling will obviously have an influence on the rate of production, but again adjustments would be made to allow the natural rhythm of the activities to be achieved.

1.9.21.8 *Material schedules*
Material schedules can be obtained easily from the Line of Balance Schedule. The amount of material required can be obtained by reading off the starts of activities at any particular time and allowing for such factors as call up times, size of loads, etc.

1.9.21.9 *Labour and plant schedules*
Labour and plant schedules can be obtained in a similar way to that used on the previous example (Fig. 1.29) simply by accumulating requirements as the starts occur and dropping them off when the first unit is complete.

1.9.21.10 *Cash flows*
Cash flows can be obtained in a similar manner to that shown in Fig. 5.2.3.

1.9.21.11 *Progressing the schedule*
Progress can be shown on the Line of Balance Schedule, e.g. assume the progress for the first thirty days is as follows:—

Excavation	completed up to and including pair 28
Concrete foundations	completed up to and including pair 19
Brickwork to D.P.C.	completed up to and including pair 13
Hardcore and backfill	completed up to and including pair 12
Site concrete	completed up to and including pair 9

Fig. 1.30

Operation	Productive operative or plant hours per pair	Min. number of productive operatives or items of plant	Time reqd. using min. gang or min. items of plant in hours
	2	3	4
Sub-structure (based on 1 pair every 7 hours)			G
Excavate oversite and trenches	14	1	14
Concrete trenches	7	1	7
Brickwork to D.P.C.	80	2	40
Hardcore fill and backfill	66	1	66
Site concrete	6	1	6
Super-structure (based on 1 pair every 14 hours)			G
Brickwork 1st and 2nd lift	332	2	166
First floor joists and roof to garage	40	2 carps. + 1 lab.	13·3
Brickwork to completion	222	2	111
Roof carcass	71	2 carps. + 1 lab.	27
Roof coverings	40	3	13·3
Flashings and glazing	54	1	54

Internal work and Finishings (based on 1 house every 7 hours)	Productive operative or plant hours per house	Min. number of productive operatives or items of plant	Time reqd. using min. gang or min. items of plant in hours
			G
Floorboards and staircase	20	2 carps. + 1 lab.	6·66
Partitions and stone fireplace	54·5	2	27·25
Carpenter 1st fix	18	2 carps. + 1 lab.	6
Electrician 1st fix	14	1	14
Plumber 1st fix	14	1	14
Heating Engineer 1st fix	30	1	30
Plastering and floor screed	158·5	2	79·25
Carpenter 2nd fix	71	2	35·5
Electrician 2nd fix	7	1	7
Plumber 2nd fix	7	1	7
Heating Engineer 2nd fix	25·5	1	25·5
Wall tiling	26·5	1	26·5
Floor finish	80	2	40
Painting and decorating	130	2	65
Snagging and final finish	14	2	7
Clean up	14	2	7

5	6	7	8	9
Total number of gangs or items of plant reqd. $g=\dfrac{G}{t}$	Number of pairs per week $R=\dfrac{35g}{G}$	Number of gangs or plant to be used on each pair (P)	Time reqd. per pair in days $T=\dfrac{G}{7P}$	Total time from start of first pair to start of last pair in days $\dfrac{5(N-1)}{R}$
2	5	1	2	29
1	5	1	1	29
6	5·25	3	1·9	27·6
10	5·3	10	0·9	27·3
1	5·8	1	0·9	24·7
$g=\dfrac{G}{14}$	$R=\dfrac{35g}{G}$	P	$T=\dfrac{G}{7P}$	$\dfrac{5(N-1)}{R}$
12	2·5	3	7·9	58
1	2·63	1	1·9	55·3
8	2·52	2	7·9	57·5
2	2·59	1	3·9	55·9
1	2·63	1	1·9	55·2
4	2·59	2	3·9	55·9
Total number of gangs or items of plant reqd. $g=\dfrac{G}{t}$	Number of houses per week $R=\dfrac{35g}{G}$	Number of gangs or plant to be used on each house P	Time reqd. per house in days $T=\dfrac{G}{7P}$	Total time from start of first house to start of last house in days $\dfrac{5(N-1)}{R}$
1	5·25	1	1·0	56·2
4	5·14	1	3·9	56·9
1	5·83	1	0·9	50·6
2	5·0	1	2·0	59·0
2	5·0	1	2·0	59·0
4	4·67	1	4·3	63·2
12	5·25	2	5·7	56·2
5	4·93	1	5·1	59·8
1	5·0	1	1·0	59·0
1	5·0	1	1·0	59·0
4	5·49	1	3·6	53·7
4	5·28	1	3·8	55·9
6	5·25	1	5·7	56·2
9	4·85	1	9·3	60·9
1	5·0	1	1·0	59·0
1	5·0	1	1·0	59·0

Fig. 1.31

Fig. 1.32

Fig. 1.33

Fig. 1.34

Fig. 1.35

Fig. 1.36

Fig. 1.37

Method 1

Shade in the activities as they are completed on each particular unit (see Fig. 1.36). An analysis of the progress to date shows that

1. Excavation is ahead of schedule.
2. Concrete foundations are on schedule.
3. Brickwork to D.P.C. is falling well behind.
4. Hardcore is approximately on schedule.
5. Site concrete is slightly in front.

It can be seen that unless brickwork to D.P.C. proceeds as scheduled originally, i.e. 5.25 pairs per week, hardcore and backfill will be delayed resulting in non-productive time for the operatives on that activity.

Method 2

Plot the actual starts and completions of each unit as they occur, see Fig. 1.37. An analysis of the progress will be as in method 1 above.

For this method it is an advantage to use job cards for each activity which are issued at the start of each activity on each unit and handed in at the completion of the activity. The cards would also be used to calculate bonus earnings.

Method 3
Plot the cumulative starts and completions for each activity at say weekly
intervals, the result will be similar to Fig. 1.37 (as method 2).

1.9.21.12 Other uses of the line of balance method
It rarely occurs that only one house type exists on a particular development. If
differing types are included in a development the line of balance method can
still be used by using a weighted average work content for each activity. This
will not apply of course if the differences are substantial.

The method can also be used in other situations where repetitive work occurs,
such as finishings in multi-storey buildings.

1.9.21.13 Learning curves
It has been established that as operatives become experienced on a particular
activity, the time required for performing it is reduced. This reduction can be
as much as fourteen per cent each time the number of units they complete is
doubled. This has not been allowed for in the Line of Balance Schedule, but it
could be incorporated.

1.10 SITE LAYOUT

The physical factors of the site often affect the method and sequence adopted
in the construction programme, so the programme and the site layout are
usually prepared together.

1.10.1 Factors affecting site layout
Before attempting to construct a site layout drawing, a list should be
compiled of all accommodation, plant and material storage areas required.
The areas on site which are available to the contractor should then be
investigated to assist in determining the overall arrangement of the items
mentioned above, i.e. where would the staff accommodation or the major
items of plant be best situated? This should be considered in overall terms at
an early stage, followed by a consideration of the more detailed aspects of the
site layout.

For this work a detailed site drawing is necessary to a scale of at least 1 : 100,
showing all foundations, drains, etc. The most suitable drawings are usually the
site general arrangement, the ground floor plan and any other drawing that gives
the full extent of the site and shows all obstructions. As changes will be made
frequently in arriving at the site layout, two common methods may be used to
facilitate alterations:

1. The scale drawing can be overlayed with a sheet of clear plastic on which
the planner will draw his tentative layout with a wax crayon that can easily be
erased if parts of the layout are later found to be unsuitable..

2. Silhouettes or models of all items of accommodation, plant and storage
areas are produced to the same scale as the layout drawing and then placed on
the scale drawing in the areas set aside for particular activities; they can now be
moved around until the required layout is achieved. This method is a work study
technique which is very useful in obtaining the best layout.

Temporary roads may or may not be necessary, depending upon the type of
sub-soil experienced on site, but when they are used they should be planned to
serve all major items of plant and material storage compounds; hardcore or
railway sleepers may be employed as construction material. In some cases the

site service roads may be completed before building work commences, in which case they could be used to provide access about the site.

A number of factors can influence the location of accommodation, plant, storage areas and temporary roads, and these are considered in the following example (based on the amenity centre).

1.10.2 Site layout for amenity centre (Fig. 1.38)

1.10.2.1 Access to the site

Ideally there should be an entrance and an exit to the site, as this has the effect of stimulating the flow of traffic. In this case the site is situated on a major road and the police would not allow the contractor access from that side: in view of this restriction and the shape and size of the site it is impracticable to provide more than one opening, which will have to serve as both entrance and exit.

A cross-over must be provided adjacent to this entrance as the footpath was not designed to take heavy vehicles, and the kerb stones and pavings on the footpath must be removed in the area of the cross-over and replaced by a temporary sleepered cross-over. As the access is not visible from the main road and is to be via the works entrance, it must be clearly signposted to direct lorries delivering materials. Visitors and personnel will enter the site via the gate adjacent to the wages office.

1.10.2.2 Hoarding

A licence must be obtained from the local authority to erect the hoarding adjacent to the main road and extending 0.6 m onto the footpath so that the boundary wall can be built. Permission has been obtained from the works manager to use the footpath at the rear of the site for extra storage space and it has been agreed that a temporary footpath and hoarding need not be provided.

The gate in the hoarding will open inwards so as not to cause obstruction in the road—a useful alternative would be the provision of sliding doors. As the main access to the site is via the works entrance (which is manned by the works police) it is considered that additional security is not necessary.

1.10.2.3 Temporary roads

Due to the shape of the site there will be only one temporary road, as shown in the plan. This will be constructed of hardcore laid at a level which is suitable for it to become part of the car park sub-base at a later date.

1.10.2.4 Plant

1. Tower crane

The crane selected has a lifting capacity of 500 kg at a maximum radius of 21 m and has been selected because with this jib length it will cover all the building and stacking grounds situated in the position indicated and will lift the required loads. As the crane is to remain static, provision must be made for anchoring the base of the mast by casting a concrete foundation block.

2. Concrete mixer

It has been decided to mix the concrete on the site because of delivery difficulties and varying quantity demand which ready mixed would have had difficulty in coping with. The size of the mixers selected is a 300/200 litre, which is electrically powered and provided with a cement silo of 20 tonnes capacity mounted directly above it, thus reducing the space required. The aggregate storage areas will be situated adjacent to the road for ease of access, the aggregates themselves will

Fig. 1.38

be kept apart by sleepered division walls. The mixer hopper is to be charged by means of a hand scraper drawn by an electrically operated winch situated on the mixer.

The concrete will be transported from the mixer by the tower crane using a roll-over skip.

3. Power bender

This will be situated in the area set aside for reinforcing. It will be located on a concrete hard standing and is covered over with a tarpaulin on a scaffold tube frame to provide a waterproof working area.

A hand operated cutter will also be provided in the covered area.

4. Circular saw

This will be situated in the area set aside for formwork, and requires a hard standing and weather sheeting in a similar manner to the power bender.

5. Hoist

A hoist will replace the tower crane when the frame is completed and the finishings are being carried out. It is to be situated in a central position, thus supplying all the building.

6. 150/100 litre mixer.

The mixers are placed in the areas shown, paving first being removed to avoid damaging it. The mixers and their related material storage areas will be located adjacent to the hoist to facilitate the delivery of mixed material into the building.

7. Scaffolding

The scaffolding is of the independent type and will be used: (i) to give access to columns and beams when concreting, (ii) to give access to brickwork, and (iii) when fixing curtain walling. With this in mind the platforms have been made 1 m wide.

A ramp will be constructed to provide access to all platform levels.

1.10.2.5 Materials

It is good practice to mark out all materials storage areas on site and then ensure that they are strictly adhered to.

1. Reinforcing steel

All steel will be cut and bent on site and straight bars will be stored in scaffold tube racks to cut down the space required.

It is essential on such a confined site that the programme for delivery, i.e. a floor at a time, is strictly complied with. The area set aside for reinforcing steel which has been cut and bent could be further subdivided into areas which would have reference numbers corresponding to pages of the bending schedule, e.g. beams 1 to 5 for bending schedule pages 22 to 23. These areas and reference numbers would obviously be altered as and when the steel changed from floor to floor.

2. Formwork

All formwork is to be produced on site in the area provided. A joiners' shop will be provided and will consist of a scaffold tube frame covered with a tarpaulin and will be located on a concrete hardstanding.

3. Bricks and blocks

Bricks are to be delivered in packaged form of 200 bricks per pack, and the area set aside must be within reach of the tower crane (giving consideration to the weight involved). This material will be off-loaded by means of the lorries' own cranes. Blocks are to be delivered on palets and unloaded in a similar manner to the bricks.

4. Windows and curtain walling

All the glass involved will be stored inside the building. The windows and curtain walling units will be delivered a floor at a time and stored in the areas shown, special care being taken with the storage, i.e. the windows must be stacked vertically on a firm base or in a scaffold tube rack.

5. Cast iron drains and stacks

These will be ordered from the local builders' merchants as required to obviate storage on site.

All other materials will be stored within the building when the shell is erected, except for roof covering materials which are to be stored on the roof.

1.10.2.6 Accommodation

1. Location

The area to be used for the accommodation has been selected for a number of reasons:

(a)   It is possible to have all accommodation in one area, which helps communication between personnel.

(b)   It is in an area which is likely to be less noisy than the others, i.e. away from joiners shop, mixer, etc.

(c)   It efficiently utilizes an area which would be difficult to employ for any other purpose as it has poor access.

(d)   The wages and time office can easily be located against the door provided for the access of workmen.

(e)   People visiting the site manager can conveniently gain access to site and readily locate his office. The same point applies equally to people seeking the clerk of works.

(f)   As this is a garden area which is a separate contract, it is possible to complete the site works without moving accommodation, i.e. the accommodation is in one position throughout the project.

2. Construction and fittings

The accommodation will be of sectional timber construction with weather board walls and felted roof. The inside is to be lined with insulation boarding to cut down heat loss.

All compartments will be provided with lighting in the form of fluorescent strips and electric fires.

3. Accommodation provided

(a)   Site manager, assistant site manager, site clerk, and cost surveyor

These site staff will be kept together in one office complex for ease of communication. A first-aid box will be located in the assistant site manager's office.

(b)   Clerk-of-works

The clerk-of-works' office will be situated as close to the contractor's office as possible for ease of communication.

(c)   Mess room

This will be provided with a sufficient number of benches and tables for men to take their meals, water for drinking, and facilities for boiling water. It will be situated adjacent to the trades foremen's accommodation to reduce the amount of prolonged tea break.

(d)   Drying and locker room

Accommodation for hanging up off-site and/or on-site clothing to dry is to be provided at the end of the mess room. Lockers will also be available for workmen's personal effects.

(e)   Water closets

These should be situated away from the messroom and administrative area, but not too far away to cause inconvenience.

The facility is in the form of a mobile amenity unit having two w.c.'s, two washbasins, a urinal, and an instantaneous hot water heater in one compartment, whilst another compartment is provided with one w.c., one wash basin and an instantaneous hot water heater (this latter compartment being for the use of the staff and the clerk of works).

The amenity unit will be connected to an existing manhole and existing water and electrical service as shown.

1.10.2.7 Services

1. Water

Water will be supplied to all mixer points (terminating at a standpipe), the messroom, and the mobile amenity unit.

2 Electricity

There will be two supplies, one being a 415-volt three-phase for the plant, whilst the other is 240 volts for the heating and lighting in the accommodation. The supplies will be taken to weather-proof boxes containing distribution boards, cut-outs and meters that are located on the side of the store and the clerk-of-work's office.

The 240-volt supply would be stepped down to 110 volts or 50 volts by means of a site transformer for powering small tools used on site. The transformer can also be housed in a weather-proof box or alternatively a small weather-proof transformer can be used out on the site.

3. Telephones

These would be connected as soon as the site accommodation is completed. Telephones would be provided for the site manager, the site clerk, and the clerk-of-works.

1.11   MULTI-PROJECT PLANNING

1.11.1 Introduction

With the planning techniques used so far in this book each project has been considered in isolation. Little attention has been paid to the demand each project has placed on the total resources of the firm, providing that the maximum amount of any resource has not been exceeded. In practice many projects will be proceding at the same time and it is obvious that the total available resources and the demand for such resources by all current and future known projects should be considered when drawing up programmes for new projects. The following outlines a method which enables planning of a number of projects to take place within the total available resources of the firm or section of the firm and facilitates the coordination of a number of projects on the one chart.

1.11.2 Method of presentation

Using this planning technique all operations which follow each other within a particular project are recorded on the same line of the chart. This saves considerable space on the chart thus allowing more projects to be presented. For clarity, each block on the chart can be coloured using a particular colour for each trade and arrows of the same colour can be used to indicate the transfer of men from one project to another. If presentation is in black and white then different types of shading would be used.

1.11.3 Exceeding available resources

If the total available resources within the firm must be exceeded then adequate advanced warning is given that further recruitment of labour or hire of plant required. Occasionally, in certain areas of the country, the demand for tradesmen such as carpenters exceeds supply and in such circumstances this planning technique is invaluable in making the best use of available tradesmen and providing them with continuity of employment, thus retaining their services.

1.11.4 Procedure

As each project is awarded to the firm it will be integrated with the previously planned projects which are all displayed on a master chart. As stated earlier, care would be taken when deciding on the timing of operations to see that the maximum available resources are not exceeded. When the new project has been added to the master chart, the programme for each projects manager and trade foreman can be taken off and presented separately. It is particularly important that the trades foreman should have a copy of this programme as they may be involved in working for more than one projects manager as well as different site managers under a particular project manager.

1.11.5 Programme for four projects (see Fig. 1.39)

As an example of the use of this technique four projects are considered. Project 1 being shops and offices, project 2 being a gymnasium, project 3 is the Amenity Centre and project 4 is the substructure to a building. The four projects are under the supervision of one projects manager.

1.11.5.1 Resource levels

The maximum level of each labour resource is as follows:—

> Carpenters 10
> Plumbers 2 + 2
> Electricians 2 + 2
> Steelfixers 8
> Bricklayers 8 + 4
> Plasterers 4 + 2
> Tilers 6 + 3
> Painters 8
> Scaffolders 2
> General labourers   no max. set

1.11.5.2 Projects 1 and 2

The particular projects manager was given projects 1 and 2 at the same time. They start within two weeks of each other. Both these projects are planned and co-ordinated so as not to exceed the resources under the control of the projects manager.

1.11.5.3 Continuity of work in projects 1 and 2

To illustrate how continuity of work is achieved in projects 1 and 2 the trade of bricklayer will be used. In project 1, half way through week 2, two bricklayers are required for underpinning work, this work being almost complete by the end of week 5 when these two bricklayers are moved to project 2 for one week. They are then transferred back to project 1 to complete the underpinning. At the end of week 7 the bricklayers are moved back to project 2 to carry out the brickwork in the site works. This coordination and movement of men achieves

Fig. 1.39

AY	JUNE					JULY					AUGUST				SEPTEMBER				OCTOBER				NOVEMBER		
13	20	27	3	10	17	24	1	8	15	22	29	5	12	19	26	2	9	16	23	30	7	14	21	28	4	11	18
29	30	31	32	33	34	35	36	37	38	39	40	41	42	43	44	45	46	47	48	49	50	51	52	53	54	55	56

Chart bars (labels as shown):

- 2C — CARP | 2C — 2ND FIX / CARP FINISH | →6PT→8 PT. — PAINTING | 4L — CLEAN UP SITE
- 2Pb+1 — GLAZING | 2PL+1 — PLUMBING 1ST. FIX | 2Pb+1 — PLUMBING 2ND. FIX | 2PL+2L
- 4PL+2L — SCREED | M.G. PLAST
- 2C — CARP. 2ND FIX
- 6PT. — PAINTING
- 2L EXC DRAIN | 2B+4L SITE WORKS
- 2C — CARP 1ST. FIX | 2C — CARP. 2ND. FIX
- 6b+1 BWK. CLADDING | 2PL+2L ROOF SCREED | 2PL+2L PLASTER & FLOOR SCREED | 6PT. INT. PAINTING
- 2B+1L BLOCKWORK | 4B+2L INT. PARTITIONS | 6T+3 WALL TILING | 3L CONC FOUND | 3L BACKFILLING | 2C CLEAR HUTS
- 2Pb+1 PLUMBING 1ST FIX | 2Pb+1 PLUMBING 2ND FIX | 3L EXC. | 2B+1 BWK. | 3L CLEAN & TIDY UP
- 2PT. PAINT
- 2E+2 ELECT. 1ST. FIX | 1E+1 ELECT 2ND FIX
- ION
- MWORK (4 WKS WORK ONLY) | 3L EXCAVATION | 3L | 4L CONC. BAST. FLOOR | 4L CONC. WALLS & COLS.
- 2S C&B FLR REINF | C&B WALL & COL REINF | 2S FIX BAST. FLOOR REINF. | 2S FIX WALL & COL REINF.
- 8C FWK. TO WALLS | 8C FWK. TO SLAB | 4B+2L BWK. PROT. SKIN

Resource tabulation (lower section):

0	10	10	10	10	10	10	10	10	10	10	10	2	2	2	2	2	2	2	2	2	2	2	2	2	2		
2	2	2	2	2	2	2	2	2	2	2	2	2	2	2	2	2	2	2	2	2	2						
1	1	1	1	1	1	1	1	1	1	1	1	1	1	1	1	1	1	1	1	1	1						
												2	2	2	2	1	1	1	1	1							
												2	2	2	2	1	1	1	1	1							
5	5	5	5	6	6	6	3	3	3															2	2		
				6	8	8	8	8	8	8	4	4	8	8	8	4											
										2	2	2	2	2	6	6	6	6	6								
4	4	4	4	4	4	4									8	8	8	8	8	8		6	6	6			
					6	6																					
2	2	2	2	2	2		2	2	2	2	2	2	2	2	2	2	2	2									
9	9	9	15	14	18	19	19	5	9	9	6	2	4	6	6	6	4	5	5	5	5	3	4	7	7	4	4	3	3	3	3

	1	1	1																			1	1		
1	1	1	1																						
1	1	1	1	1	1	1	1																		
				2	2	2	1	1	1	1	2	2	2	1	1		1	1	1	1	1	1		
1	1	1	1	1	1	1	1	1	1	1	1		1	1	1	1	1	1	1		2	2	
	3	3	3																	2	2	
2	2	2	2	1	1	1																
3	3	3	3	3	3	3	3	3	3	2	2	2	1	1	1	1	1	1	1	1	1	1	1
1	1	1	1													1	1		

continuity of work for this trade and at the same time keeps the demand on the resource to a minimum. Half way through week 10 it can be seen that the number of bricklayers has to be built up to eight to cope with the volume of work and that these eight are kept continuously employed on these two projects until week 23. After week 23 they are available for reallocation to another projects manager.

1.11.5.4 Project 3
At the beginning of week 17, project 3 commences. This also has to be co-ordinated with project 1 and 2 which are underway.

1.11.5.5 Continuity of work in project 3
As an example of how continuity of work is achieved here the carpenters work will be considered. During week 18, eight carpenters are working on formwork to project 1 and two carpenters are working on roofing timbers to project 2. At the start of week 19, four carpenters are transferred from project 1 to project 3 to start the formwork to the frame of the Amenity Centre, these four carpenters being joined by the remaining four from project 1 at the beginning of week 21. The two carpenters on project 2 remain there to carry out carpentry 1st. fix. By this transfer of men all three projects are provided with the resources they require without exceeding the total available number of carpenters.

1.11.5.6 Project 4
At the beginning of week 27 project 4 commences and must be coordinated with projects 1, 2, and 3. The problem here is that a relative larger number of carpenters are required and because of this the formwork to the walls will have to start half way through week 36 when the eight carpenters can be released from project 3.

1.11.5.7 Labour and plant schedules
As each new project is introduced, labour and plant schedules are drawn up to show the level of each resource week by week.

1.11.5.8 Note
1. The two scaffolders shown move between the four projects erecting and dismantling scaffolds as required, their work being planned to fit in with the scaffolding requirements on each project.

2. The plant required on each project has been omitted from the top part of the master chart for clarity in publication, but the resulting schedule has been shown at the bottom of the chart.

3. Sub-contractors have not been shown on this chart but their work has been considered and will fit in with the other operations.

2: The Critical Path Method of Planning

2.1 INTRODUCTION

The critical path method (C.P.M.) of planning is one of a number of techniques based on a network. It is particularly applicable to construction work because it lends itself to the 'one off' project.

Many other systems have been evolved using a network as their basis, such as PERT (Programme Evaluation and Review Technique) which was used in the development of the Polaris project.

Networks for small projects can be successfully analysed without the use of a computer, but for the larger projects the computer becomes necessary to save time in the analysis, re-analysis and up-dating of the network.

2.2 ADVANTAGES OF C.P.M. OVER THE BAR CHART

1. When using C.P.M., the inter-relationship of all operations is clearly shown on this network. The normal bar chart does not do this and consequently requires the dependance of one operation upon another to be remembered by the planner: this is extremely difficult with large projects, and in addition the site manager (who must carry out the work) has to be informed how dependent one operation is upon another.

2. When a delay occurs and C.P.M is being used, critical operations will stand out as requiring particular attention. When bar charts are used on a large project many operations tend to be 'crashed' unnecessarily as it is almost impossible to remember which operations are inter-dependent.

3. It is far easier for anyone taking over a partially completed project to become familiar with the progress when C.P.M. is employed.

4. When using C.P.M. it is essential to study the sequence of operations very carefully, leading to a closer understanding of the project.

5. Planning, analysing and scheduling are separated when using C.P.M., which allows a greater concentration on the planning aspect.

2.3 THE PREPARATION OF NETWORK DIAGRAMS

2.3.1 Activities and events

A network consists of activities and events. An activity is an operation, and all activities start and finish at an event, which is a point in time and may be the junction of two or more activities (Fig. 2.1). Activities take time whereas events do not, and an activity cannot be started until all the activities leading into its preceding event are completed.

Activities are represented by arrows and the sequence of the arrows represents the sequence of activities; events are normally represented by circles. To avoid misunderstanding, diagrams are arranged so that time flows from left to right.

The first stage in the preparation of network diagrams is the listing of the activities to be used. The amount of detail required in the breakdown depends on many factors, such as the stage of planning, i.e. pre-tender, pre-contract or

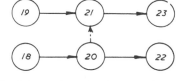

Fig. 2.1   Activity 9-10 cannot start until activities 7-9 and 8-9 are completed.

Fig. 2.2   Activity 21-23 cannot start until activities 19-21 and 18-20 are completed, but activity 20-22 can start when activity 18-20 is completed and is not dependent on activity 19-21.

Fig. 2.3.1   Figs 2.3.1, 2.3.2 and 2.3.3 represent the same situation.

Fig. 2.3.2   Fig. 2.3.3

short term. One activity in a network, drawn in the early stages, may be broken down into a number of activities at a later stage, and when a project is first being considered the erection of the building could be a single activity representing the finalization of the overall scheme. Other activities at this stage could be awaiting approvals, preparation of sketch plans, the raising of capital, and preparation of bills of quantities etc.

'Dummy' activities are sometimes necessary in a network. These do not take time to perform and are used either to make the sequence clear (Fig. 2.2) or to give a unique numbering system (Figs. 2.3.1, 2.3.2 and 2.3.3). If a 'dummy' is not used here, two activities would have the same reference, e.g. 9-10 (Fig. 2.3.1) would have the same preceding and succeeding event. This is confusing when using a computer for analysis and it is therefore essential that a unique numbering system be used (Figs. 2.3.2 and 2.3.3).

When drawing networks it is important to remember that 'off-site' activities such as the delivery of plasterboard can be critical and must therefore be included.

As each activity is considered the following questions should be asked about each:

1. Which activities must be completed before this activity can start?
2. Which other activities can proceed at the same time as this activity?
3. Which other activities cannot start until this activity is completed?

A common mistake made here is for the planner to show an activity where he thinks it will be done, rather than the earliest time at which it can be done.

2.3.2 Overlap of activities

Many activities in a building project can overlap, one activity starting before the previous activity is completed. For example, in a large concreting operation involving numerous columns, steelfixing would begin first and then formwork could be started when perhaps two columns had been completed; concreting then follows when the formwork is sufficiently advanced. To show this in a network it is necessary to split each activity.

1. Divide each activity into smaller equal portions (Fig. 2.4). This method of

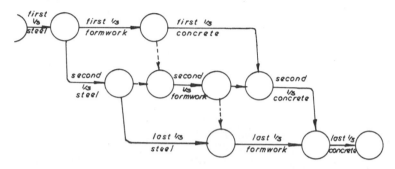

Fig. 2.4

dividing each activity into the same number of equal parts is the only method which truly portrays the situation, but if each activity is divided into many parts the diagram becomes unduly complicated and in practice one of the following methods is often used as an alternative.

2. Divide each activity into three sections; begin, continue and complete and again show the sequence in the normal way (Figs. 2.5.1 and 2.5.2).

3. Use 'ladders' and 'time restraints' (Fig. 2.6). Activities 9-11 and 11-13 are known as 'lead time' and 10-12 and 12-14 as 'lag time'; with time restraints in use, activity 9-11 represents the amount of reinforcement that must be fixed before formwork can start, activity 11-13 represents the amount of formwork to be erected before concrete can be poured, activity 10-12 represents the last of the formwork to be erected after steelfixing is completed, and activity 12-14 represents the last of the concrete to be poured after the formwork is completed.

Fig. 2.5.1

THIS CAN BE SIMPLIFIED THUS

Fig. 2.5.2

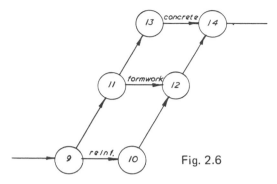

Fig. 2.6

Methods 2 and 3 will give the correct event times, but the relationship between activities may not be truly represented when the analysis is tabulated. Care is therefore necessary when using these techniques.

2.3.3 An example of the drawing of a network

Figure 2.7 shows a village hall which is attached to an existing building by means of a link corridor. The new building is to be constructed using timber portal frames with side elevations clad with brickwork and end elevations clad with timber frame units incorporating windows and doors. In this example the activities are limited to those listed below, and for simplification no overlap has been allowed.

The activities involved are: foundations, deliver frames, erect frames, brickwork up to damp-proof course, brickwork up to eaves, roof construction, roof finish, floor construction, end panels, glazing, internal partitions, electrician first fixing, joiner first fixing, deliver plasterboard, plasterer, electrician second fixing, joiner second fixing, make good plaster, floor screeds, floor finish, painting and decorating, electrician finishings, joiner finishings, rainwater goods, allow floor screed to dry, paving, drainage, clean-up and hand over.

As a further means of simplifying the network some of the activities have been grouped. For example, 'foundations' includes excavation, levelling and ramming, and concrete bed. The network based on these activities and the drawing (Fig. 2.7) is shown in Fig. 2.8, the following assumptions having been made in its construction:

1. The end panels are built off the concrete slab.
2. The internal partitions cannot be built until the external brickwork and end panels are complete.
3. The building must be weatherproof before joiner and electrician first fixings are carried out.

The sequence of activities shown in Fig. 2.8 is not unique and other planners may draw this network slightly differently.

2.4   ANALYSIS OF NETWORK DIAGRAMS

2.4.1 Activity duration

Activity durations (shown on the arrows) can be in days, weeks or months, depending on how detailed the network is to be. The durations for contractors activities would be arrived at by using past records of outputs, synthetic times based on work study, etc. The gang sizes and plant to be used would be the ones which give optimum performance at this stage irrespective of the requirements of other activities. Sub-contractors and suppliers should be approached in order to obtain realistic durations for their activities.

VILLAGE HALL

NOTES
Gable walls to be framed up in softwood &
covered with horizontal cedar boarding
Window units in afromosia

WEST ELEVATION

EAST ELEVATION

SOUTH ELEVATION

existing drain

new manhole

new s.w. drain

paved area

link to existing building
already built

OFFICE

CLOAKS

HALL

STAGE

12.900 m

22.600 m

4.300 m

PLAN

N

Fig. 2.7

SECTION A-A

3 layers felt on cement slurry
on 3 reinforced woodwool
slabs

225×75 purlins

plasterboard ceiling on 75×50 noggings

laminated timber portals at
3.600 m cts

12.300 m span

300 cavity walls

carlite plaster
on walls

10 m.s shoe

h.w. strip on 75×50 battens

150 concrete 1:2:4
150 hardcore

strip foundation 1:3:6 conc.

isolated foundation 1:3:6 conc.

5.900 m

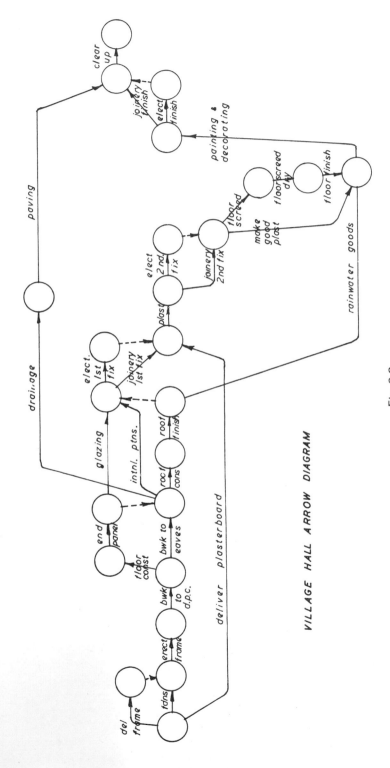

VILLAGE HALL ARROW DIAGRAM

Fig. 2.8

2.4.2 Project duration

The project duration is the minimum time in which it can be completed with the activity times assigned to it.

The duration of the project will be determined by the longest path through the network. This is known as the critical path.

If the time required for any activity is affected, this will automatically affect the project duration.

2.4.3 Method of Analysis (Fig. 2.9)

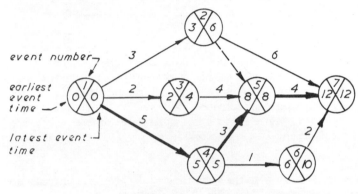

Fig. 2.9

2.4.3.1 Calculate the earliest event times by working through the diagram and selecting the longest path.

Earliest time for event 1 is day 0:

Therefore the earliest time for event 2 is 0 + 3 = 3 weeks

for event 3 is 0 + 2 = 2 weeks

and for event 4 is 0 + 5 = 5 weeks

Event 5 has three activities leading into it, and the earliest time for event 5 is determined by the longest path. The alternative paths give the following results:

event 2 to event 5 would give 3 + 0 = 3 weeks

event 3 to event 5 would give 2 + 4 = 6 weeks

event 4 to event 5 would give 5 + 3 = 8 weeks

The longest path to event 5 is therefore from event 4 to event 5 which means the earliest time for event 5 is 8 weeks.

The remainder of the earliest event times are calculated in this way, and the earliest time for the final event gives the project duration.

2.4.3.2 Calculate the latest event times by working backwards through the diagram again, selecting the longest path.

The latest event time is the latest time an event can be completed without affecting the project duration, so that the latest time for the final event is the same as its earliest time (the project duration).

Therefore the latest time for event 7 is week 12

for event 6 is 12 − 2 = 10 weeks

and for event 5 is 12 − 4 = 8 weeks

Event 4 has two activities leaving it, so the latest time for event 4 is determined by the longest path. The alternative paths give the following results:

event 5 to event 4 would give   8 − 3 = 5 weeks
event 6 to event 4 would give 10 − 1 = 9 weeks

The longest path to event 4 is therefore from event 5 to event 4, which means that the latest time for event 4 is 5 weeks.

Fig. 2.10

Critical events have earliest times equal to latest times. The *critical path* passes through these events, along the activities whose duration is equal to the difference between the preceding and succeeding event times (Fig. 2.10). The *critical path* passes along activities 25-26 and 26-27 but 25-27 is not critical as activity duration of three weeks is not equal to the difference between the times for events 25 and 27, i.e. 10 weeks.

2.4.4 Float

Float is the name given to the spare time available on non-critical activities. There are various types of float (Fig. 2.11) as follows.

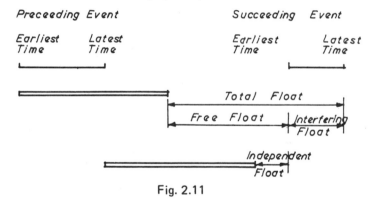

Fig. 2.11

2.4.4.1 Total float
Total float is the total amount of spare time available in an activity. It is calculated by: latest succeeding event time— earliest preceding event time — activity duration. It should be noted that total float can be the amount of float in a chain of activities.

2.4.4.2 Free float
Free float is a part of total float and represents the amount of spare time which can be used without affecting subsequent activities, providing the activity starts at its earliest time. It is calculated by: earliest succeeding event time − earliest preceding event time − activity duration.

2.4.4.3 Interfering float
Interfering float is the amount of spare time available. which, if used, will affect subsequent activities, and may be calculated by: total float − free float.

2.4.4.4 Independent float

Independent float is the amount of spare time available which can be used without affecting any succeeding activity and which cannot be affected by any preceding activity. The calculation is: earliest succeeding event time – latest preceding event time – activity duration.

2.4.5 Significance of the various types of float

If additional resources were required on a critical activity they could be obtained from:

1. Activities with independent float without affecting the rest of the network at all.

2. Activities with free float without affecting the float of subsequent activities.

3. Activities with interfering float only, which will affect the float of previous and subsequent activities.

If only the total float is calculated, this may include independent, free and interfering float.

2.4.6 Tabular presentation of networks

The analysis of the network can be presented in tabular form, and when computers are used in analysis this is the way in which the results will be presented. The table shown in Fig. 2.12 presents the information in terms of activities in numerical order, but the sequence can be varied in many ways (e.g. by arranging in the order of least total float). The table can also be used as a means of controlling the project.

Fig. 2.12

Activity number	Duration	Earliest		Latest		Total float
		Start	Finish	Start	Finish	
1.2	3	0	3	3	6	3
1.3	2	0	2	2	4	2
1.4	5	0	5	0	5	0
2.5	0	3	3	8	8	5
2.7	6	3	9	6	12	3
3.5	4	2	6	4	8	2
4.5	3	5	8	5	8	0
4.6	1	5	6	9	10	4
5.7	4	8	12	8	12	0
6.7	2	6	8	10	12	4

2.4.7 Method of analysis

1. List the activities in numerical order.

2. Fill in the earliest start time, which is the earliest time for the preceding event (obtainable from the network; Fig. 2.9).

3. Calculate the earliest finish time, which is the earliest start time plus activity duration.

4. Fill in the latest finish time, which is the latest time for the succeeding event.

5. Calculate the latest start time, which is the latest finish time minus the activity duration.

6. Calculate the total float, which is either the latest start time minus the earliest start time or alternatively the latest finish time minus the earliest finish time. Both methods give the same result.

It should be noted that critical activities have no float.

2.4.8 Analysis of the village hall project

Figure 2.13 shows the activity durations and resources for the village hall in Fig. 2.7. The analysed network for this project is shown in Fig. 2.14 and the completed table of results with total float in Fig. 2.15.

Fig. 2.13

Activity	Man-hours	Plant-hours	Resources	Time required (days)
Foundations				
Excavate oversite strip	} 40	20	J.C.B. 4 + operator + 2 labourers	} 5
Excavate for foundations				
Concrete foundations	160	20	8 labourers 400/300 litre mixer	
Frame				
Delivery				10
Erect	16	8	Crane + operator + 2 labourers	1
Brickwork to D.P.C.	96	16	4 bricklayers 2 labourers 150/100 litre mixer	2
Floor construction				
Hardcore	208	24	8 labourers 400/300 litre mixer	3
Concrete				
Brickwork to eaves	528	88	4 bricklayers 2 labourers 150/100 litre mixer	11
End panels	208		2 joiners	13
Glazing	112		Plumber and mate	7
Roof construction				
Carcass	240		2 joiners	15
Woodwool				
Internal partitions	96	16	4 bricklayers 2 labourers 150/100 litre mixer	2

Fig. 2.13 continued

Activity	Man-hours	Plant-hours	Resources	Time required (days)
Drainage				
Excavation and backfill	136	6	8 labourers 400/300 litre mixer	4
Concrete bed				
Lay drains	32	16	Bricklayer and mate 150/100 litre mixer	
Manholes				
Roof finish	144		2 roofers	9
Rainwater goods	32		Plumber and mate	2
Electrician first fixing	112		Electrician and mate	7
Joiner first fixing	96		2 joiners	6
Deliver plasterboard				70
Plastering	410	80	4 plasterers 2 labourers 150/100 litre mixer	10
Electrician second fixing	48		Electrician and mate	3
Joinery second fixing	80		2 joiners	5
Make good plaster	8		Plasterer	1
Floor screed	160	32	4 plasterers 1 labourer 150/100 litre mixer	4
Floor screed dry				10
Floor finish	192		2 joiners	12
Painting and decorating	256		4 painters	8
Paving	144		Bricklayer and mate	9
Electrical finishing	16		Electrician and mate	1
Joinery finishing	48		2 joiners	3
Clean and hand over	40		1 labourer	5

Fig. 2.14

VILLAGE HALL

Fig. 2.15

Activity	Activity number	Duration	Earliest		Latest		Total float
			Start	Finish	Start	Finish	
Deliver frame	1-2	10	0	10	10	20	10
Foundations	1-3	5	0	5	15	20	15
Deliver plasterboard	1-13	70	0	70	0	70	0
Dummy	2-3	0	10	10	20	20	10
Erect frame	3-4	1	10	11	20	21	10
Brickwork to D.P.C.	4-5	2	11	13	21	23	10
Floor construction	5-6	3	13	16	23	26	10
Brickwork to eaves	5-8	11	13	24	28	39	15
End panels	6-7	13	16	29	26	39	10
Dummy	7-8	0	29	29	39	39	10
Glazing	7-11	7	29	36	56	63	27
Roof construction	8-9	15	29	44	39	54	10
Internal partitions	8-11	2	29	31	61	63	32
Drainage	8-21	4	29	33	109	113	80
Roof finish	9-10	9	44	53	54	63	10
Dummy	10-11	0	53	53	63	63	10
Rainwater goods	10-20	2	53	55	109	111	56
Electrician first fixing	11-12	7	53	60	63	70	10
Joiner first fixing	11-13	6	53	59	64	70	11
Dummy	12-13	0	60	60	70	70	10
Plastering	13-14	10	70	80	70	80	0
Electrician second fixing	14-15	3	80	83	82	85	2
Joiner second fixing	14-16	5	80	85	80	85	0
Dummy	15-16	0	83	83	85	85	2
Make good plaster	16-20	1	85	86	110	111	25
Floor screed	16-18	4	85	89	85	89	0
Floor screed dry	18-19	10	89	99	89	99	0
Floor finish	19-20	12	99	111	99	111	0
Painting and decorating	20-22	8	111	119	111	119	0
Paving	21-24	9	33	42	113	122	80
Electrician finish	22-23	1	119	120	121	122	2
Joiner finish	22-24	3	119	122	119	122	0
Dummy	23-24	0	120	120	122	122	2
Clean and hand over	24-25	5	122	127	122	127	0

2.5   REDUCING THE PROJECT DURATION

2.5.1 Introduction

The first analysis of a network will often result in an excessive duration. The critical path method is extremely useful in these circumstances as it highlights those activities which must be examined in order to reduce the project duration. The aim will be to reduce the time required for those activities which add least cost to the project, and least cost methods have been developed for this process. Detailed consideration is beyond the scope of this text but a simple example illustrating the use of such methods is given at the end of this section.

A study of the network will often indicate which activities can be reduced at least cost when reducing the project duration by a certain amount. In practice decision rules can be used to help select the best course of action, and the most effective procedure is to increase the size of the gang for the trade which has the highest total time on the critical path. This will have an effect on many critical activities and sometimes cuts down the project duration considerably. Very often gang sizes can be increased at little or no extra cost, particularly when the reduction in overheads due to the reduction in the project duration is taken into account.

2.5.2 Example of the reduction of project duration

It is necessary to place restrictions on the example to keep it relatively simple.

Figure 2.16 represents a small building project, the duration of each activity and the labour required being given in Fig. 2.17. From this information the critical path and the project duration can be found. For simplicity it is assumed that the operatives are totally interchangeable, although the number of operatives on the operations must be as given.

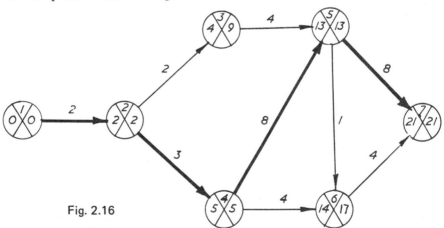

Fig. 2.16

2.5.2.1 General information

All the activities are to be performed by casual labour working a five-day week. The labour force can be recruited as required and employed on a day-to-day basis at £4 per day, with an engagement charge of £6 per person for each new appointment or re-appointment. Overtime can be applied but must only be worked in periods of three days or multiples of three days; overtime working will account for an increase in output of $33\frac{1}{3}$ per cent, the increased cost of overtime working being calculated at a rate of £3 per day per operative.

Site overheads are calculated at the rate of £35 per week, and the employer requires the project to be completed in 19 weeks in the first instance.

Fig. 2.17

Activity	Duration	Gang size	Earliest		Latest		Total float
			Start	Finish	Start	Finish	
1.2	2	10	0	2	0	2	0
2.3	2	2	2	4	7	9	5
2.4	3	5	2	5	2	5	0
3.5	4	6	4	8	9	13	5
4.5	8	3	5	13	5	13	0
4.6	4	4	5	9	13	17	8
5.6	1	2	13	14	16	17	3
5.7	8	8	13	21	13	21	0
6.7	4	4	14	18	17	21	3

2.5.2.2 Solution

It can be seen that the diagram as analysed gives a project duration of 21 weeks. As the employer requires the project completed in 19 weeks, the duration must be reduced by 2 weeks.

Figure 2.18 shows the network converted to a bar chart with all the activities at their earliest times. The cost of the project as shown in this fig. would be:

Engagements	25 at £6	£150
Labour costs	185 man-weeks × 5 × £4	£3700
Overheads	21 weeks at £35	£735
	Cost	£4585

To decrease the project duration by 2 weeks, overtime will have to be worked and in order to keep the extra cost to a minimum it will be worked on the critical activity which will cost the least, i.e. the activity with the least labour (in this case 4-5). If overtime is worked for 30 days, 40 days work will be done (due to a $33\frac{1}{3}$ per cent increase in output) and this will reduce the contract duration by two weeks.

ACT.	1	2	3	4	5	6	7	8	9	10	11	12	13	14	15	16	17	18	19	20	21
1-2	10	10																			
2-3			2	2																	
2-4			3	3	3																
3-5					6	6	6	6													
4-5						3	3	3	3	3	3	3	3								
4-6						4	4	4	4												
5-6														2							
5-7														8	8	8	8	8	8	8	8
6-7															4	4	4	4			
	10	10	7	7	11	13	13	13	7	3	3	3	3	10	12	12	12	12	8	8	8

Fig. 2.18

The ammended diagram in bar chart form will now be as shown in Fig. 2.19 and the project cost will be:

Engagements	25 at £6	£150
Labour costs	179 man-weeks × 5 × £4	£3580
Overtime	30 days × 3 men × £3	£270
Overheads	19 weeks at £35	£665
		Cost £4665

Therefore the least extra cost = £4665 − £4585 = £80

Fig. 2.19

Note

Resource allocation is not considered here, so all activities are left at earliest times.

By allocating the resources, the number of engagements and re-engagements could be reduced.

2.6   RESOURCE ALLOCATION

2.6.1 Introduction

Resource allocation involves the levelling or smoothing of the resource requirements, the aim being to keep fluctuating requirements for labour and plant to a minimum. In doing this the plan is transformed into a work schedule.

When drawing the arrow diagram and estimating the activity times, the availability of resources for the different activities was not limited, and it was therefore assumed that they would be available as required. In some cases, however, the same resources may be required on different activities which fall on parallel paths in the arrow diagram.

When allocating resources and smoothing requirements, activities can be moved within their float, but if after doing this demand for resources still exceeds availability the project duration will have to be lengthened or some alternative method considered, e.g. working overtime.

2.6.2 Example of resource allocation

Assume that in the network in Fig. 2.20 all the men are capable of being employed on all the activities and that each activity requires the number of men shown in Fig. 2.21.

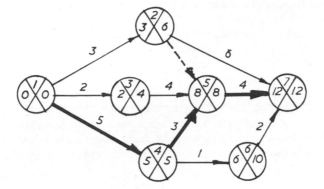

Fig. 2.20

Fig. 2.21

Activity	Number of men
1-2	3
1-3	4
1-4	6
2-7	5
3-5	2
4-5	3
4-6	1
5-7	5
6-7	2

The bar chart showing all the activities at their earliest times is drawn first as in Fig. 2.22, from which it can be seen that with the activities at their earliest times the labour requirement is fluctuating. As the critical activities cannot be moved they are drawn in first, and other activities are then moved within their float period to bring about a levelling of the resource requirement (Fig. 2.23). It must be appreciated that in allocating resources some float is used up.

The term scheduling is used in C.P.M. for the process of giving dates to the activities when the resources have been allocated.

2.7 CONTROL WITH C.P.M.

2.7.1 Introduction

Because C.P.M. draws attention to those activities which affect the project duration, it is an excellent basis for control. More attention can be given to those activities which are critical and near critical, although this does not mean that activities with a reasonable amount of float can be ignored altogether or they themselves may then become critical.

For site use the arrow diagram is usually converted into a bar chart, which is used for short term control and shows operations at scheduled times; progress is then marked onto this and the information from the bar chart can be transferred back onto the arrow diagram for re-analysis or up-dating of the network for the whole project.

An alternative method of presentation is to draw the network on a time scale showing activities at scheduled times and to mark the progress directly onto this. (see section 2.10). It is possible to use the network in its original form and to enter the actual durations on the network for re-analysis: if the results are tabulated and the actual durations are entered on the schedule this will form a method of control.

Re-analysis of the network should be carried out at regular intervals (say fortnightly) so that any change in critical activities is brought to the attention of the management and corrections can be made in reasonable time.

If the arrow diagram is used to show progress, and it is not drawn on a time scale, elasticated string can be used to give an impression of progress.

ACT.	1	2	3	4	5	6	7	8	9	10	11	12
1-2	3	3	3									
1-3	4	4										
1-4	6	6	6	6	6		CRITICAL					
2-7					5	5	5	5	5	5		
3-5				2	2	2	2					
4-5							3	3	3	CRITICAL		
4-6							1					
5-7						CRITICAL			5	5	5	5
6-7								2	2			
	13	13	11	13	13	11	10	10	10	5	5	5

Fig. 2.22

ACT	1	2	3	4	5	6	7	8	9	10	11	12
1-2			3	3	3							
1-3	4	4										
1-4	6	6	6	6	6		CRITICAL					
2-7						5	5	5	5	5	5	
3-5				2	2	2	2					
4-5							3	3	3	CRITICAL		
4-6							1					
5-7						CRITICAL			5	5	5	5
6-7								2	2			
	10	10	11	11	11	11	10	10	10	10	10	5

Fig. 2.23

By inserting pins into activities in progress (the position being proportionate to the amount of work done on each activity) and then stretching the string from pin to pin, it would be obvious that all activities to the right of the string are still outstanding.

2.7.2 Re-analysis of a project partially completed

One method of re-analysing a project partially completed, is as follows:
1. Insert 'lead time' to represent the amount of time passed.
2. Insert duration zero on all completed activities.
3. Insert estimated time to complete partially completed activities.
4. Carry out network analysis in the normal way.

For example, assume that in the network in Fig. 2.9 a period of 5 weeks has passed and the position is:

Activities 1-2 and 1-4 are completed.
Activity 1-3 requires an estimated one week to complete.
Activity 2-7 requires an estimated five weeks to complete.
All other activities not yet started.

The re-analysed network would now appear as shown in Fig. 2.24. It can be

Fig. 2.24

seen that the critical path has changed and that the project duration is now 14 weeks. If it is necessary to complete in 12 weeks, the activities requiring attention (i.e. critical activities) are clearly shown. All activities running behind time can be demonstrated by inserting the original project duration as the latest finish time for the end event before calculating the latest finish times for the other events. From this the negative float can be calculated, which will indicate the activities that are in arrears.

A second method of showing progress on the network is to remove all completed activities, leaving only partially completed activities and those not yet started. Real time dummies are then inserted and given a value equal to the time which has passed. This replaces 1 and 2 in the preceding method, and constitutes a particularly useful alternative when the project is well advanced. For example, assume in the village hall project that after 100 days the following activities are partially completed and the estimated time necessary to complete is as shown:

Floor finish, 6 days
Rainwater goods, 1 day
All subsequent activities have not yet started.

Using the alternative method, the network will be as in Fig. 2.25.

VILLAGE HALL

in this project the progress is 5 days in front
of that planned & the new project duration
is 122 days; the critical path is the same as
before.

Fig. 2.25

2.8 COMPUTERS AND C.P.M.

For large projects it is customary to use computers to analyse and up-date the network as speed is the greatest asset and computers are the fastest, most accurate and cheapest method available.

After the network diagram has been prepared and activity times have been inserted, the events must be numbered and activities listed (with their durations) before a computer analysis can be performed. The information obtained from a computer can be in many forms, e.g. in order of activities, in order of events, and in order of the various types of float, etc. Some computers can extract sections and print them out on a separate schedule (for example sub-contractors' work).

It should be emphasized that for smaller projects, networks can be analysed and up-dated by. hand.

2.9 NETWORKS USING FLOW CHARTS

2.9.1 Introduction

Networks can be drawn in the form of flow charts as an alternative method to the more conventional arrow diagrams. When using flow charts the activities are usually represented by rectangles and the relationship between activities is shown by a line linking them, the diagram being drawn with a time flow from left to right.

Flow charts are easier to draw than arrow diagrams and are easier for site personnel to understand. If necessary the flow chart can be converted into an arrow diagram for computer analysis.

2.9.2 Methods of analysing flow charts

2.9.2.1 Method 1

To analyse the network, the activity durations are inserted in the rectangles. The earliest start and finish times are then calculated for each activity, working through the network from left to right (Fig. 2.26). The procedure is as follows:

1. First activities are given the earliest start of zero.
2. Earliest finish of first activities = zero + duration.

Fig. 2.26

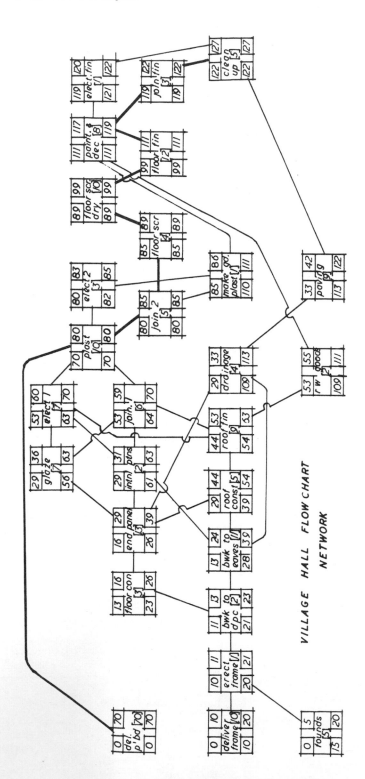

VILLAGE HALL FLOW CHART

NETWORK

Fig. 2.27 It can be seen that there are no dummy activities in this form of network: this is one of the reasons why beginners find it easier to draw

3. Each activity is dealt with in turn, the earliest start being the highest earliest finish of the activities which immediately precede it and its earliest finish time being the earliest start plus duration. The overall project time is the earliest finish time of the last activity completed.

4. Give the last activity a latest finish time equal to the overall project time.

5. Work backwards through the diagram taking each activity in turn, the latest finish time being the smallest of the latest start times of activities which immediately follow it.

6. Calculate the total float by deducting the earliest start time from the latest start time.

Figure 2.27 shows the village hall project using this method.

2.9.2.2 Method 2: using the method of potentials

In this method the activity durations are applied to the links between activities. The duration represents the time required between the end of the preceding activity and the start of the succeeding activity (Fig. 2.28).

Fig. 2.28

It is possible to show the overlap of activities when using both methods, but it is easier to do so when using the method of potentials. Furthermore, when using this method it is possible to analyse a network using a standard C.P.M. computer programme.

A disadvantage of flow charts is that they cannot be presented on a time scale, and if this is necessary then the conventional arrow diagram would have to be used. Networks in flow chart form are now becoming popular in the U.S.A. and can be analysed by computer when using the method of potentials.

2.10   TIME SCALE PRESENTATION OF C.P.M.

2.10.1 Introduction

Networks can be drawn in the form of a bar chart or time bar diagram in which case the activities are normally drawn as parallel lines to a suitable time scale. This form of presentation defines the time limitations of each activity and the relationship between activities.

The first stage when constructing an activity progress chart is to compile an activity sheet which records the sequence in which the work is to be carried out and the duration of the individual activities. The information on this sheet is then transferred to the time bar diagram for diagrammatic presentation.

2.10.2 Activity sheet (see Fig. 2.29) (taken from village hall project Fig. 2.15.)

This is a list of all the activities, their durations and any restraints involved in the project.

The first step is to list all the activities involved in the project, preferably in the order in which they are to be carried out, but this is not essential; these activities are then numbered. The next step is to provide the information regarding the relationship between activities by means of a sequence of numbers alongside the activity number. These numbers indicate which activities must be completed before the activity under consideration can be started. For example in Fig. 2.29 in the sequence column for activity 1, are the figures 0;1 this indicates that there are no other activities to be completed before activity 1 can start. Another example is activity 5, in the sequence column are the figures 1, 2;5 which indicates that activity 1 (deliver frame) and activity 2 (foundations) must be completed before activity 5 (erect frame) can start. The other column records the duration of each activity which would be arrived at via the work content and resources applied to it. This information can now be recorded diagrammatically in time bar diagram form and the critical path ascertained.

2.10.3 Time bar diagram (see Fig. 2.30)

In this chart, each activity is represented by a straight line drawn to a suitable scale. The length of each line is proportional to the duration of the activity being represented. The chart can either be drawn vertically or horizontally. If vertically then the chart is started at the top of the sheet, if horizontally then the chart is drawn from left to right as is common with most forms of bar chart.

Assuming that the chart is to be drawn horizontally then the first step is to set up a suitable time scale on the horizontal axis and then to erect a perpendicular at time zero, this line being called the 'start' line. The job lines are now drawn on the chart in the sequence indicated in the activity sheet (see Fig. 2.29), each activity being given its respective number for identification purposes. A short vertical line is introduced at the end of each activity to indicate the end of that particular activity and to separate it from the next in sequence. The relationship with activities in other parts of the chart is indicated by a longer vertical line. If there is a place in the project where a number of activities must be completed so that another activity can be started then the vertical line is drawn at the end of the activity line which is latest in time and produced to connect all activities involved. These vertical lines can also be extended when a number of activities are to be started at the same time. The vertical lines are usually called 'time lines'. The float on activities is indicated by a dotted line at the end of the activity line. This dotted line is extended to the next time line indicated on the activity sheet. The critical path is indicated by a continuous set of activities which can be arranged to fall in one straight line if required but this is not essential.

When the last activity has been drawn on the chart, another long vertical line is drawn at this point. This is called the finish line and represents the completion of the project. The project duration is therefore the dimension between the start and finish lines.

The advantage of this chart is that it shows clearly the relationship that exists between all activities in the project, the sequence of all activities and which are concurrent.

Following these basic principles, the construction of the time bar diagram, Fig. 2.30 is therefore as follows:— Having calibrated the horizontal axis (time scale) and drawn the start line, the next step is to draw on activity 1 commencing

Fig. 2.29   ACTIVITY SHEET — VILLAGE HALL

Activity Number	Activity	Activity Sequence	Activity Duration (days).
1	Deliver frame	0; 1	10
2	Foundations	0; 2	5
3	Deliver plasterboard	0; 3	70
4	Erect frame	1, 2; 4	1
5	Brickwork to d.p.c.	4; 5	2
6	Floor construction	5; 6	3
7	Brickwork to eaves	5; 7	11
8	End panels	6; 8	13
9	Glazing	8; 9	7
10	Roof construction	7, 8; 10	15
11	Internal partitions	7, 8; 11	2
12	Drainage	7, 8; 12	4
13	Roof finish	10; 13	9
14	Rainwater goods	13; 14	2
15	Electrician first fixing	9, 11, 13; 15	7
16	Joiner first fixing	9, 11, 13; 16	6
17	Plastering	3, 16, 15; 17	10
18	Electrician second fixing	17; 18	3
19	Joiner second fixing	17; 19	5
20	Make good plaster	18, 19; 20	1
21	Floor screed	18, 19; 21	4
22	Floor screed dry	21; 22	10
23	Floor finish	22; 23	12
24	Painting and decorating	14, 20, 23; 24	8
25	Paving	12; 25	9
26	Electrician finish	24; 26	1
27	Joiner finish	24; 27	3
28	Clean and handover	25, 27, 26; 28	5

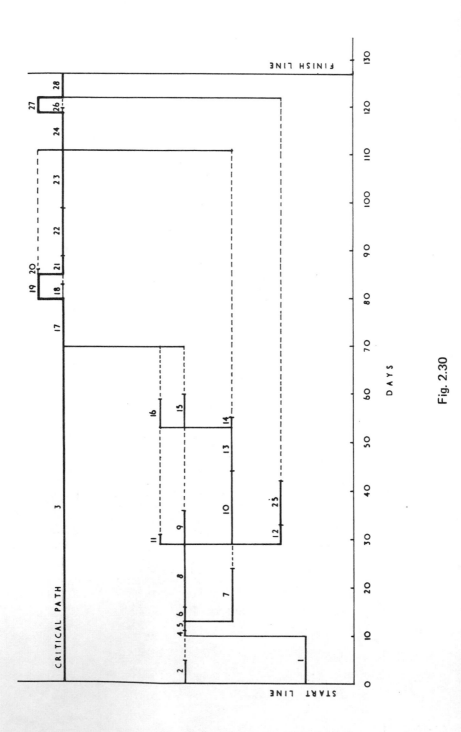

Fig. 2.30

the line at the start line (time zero) and extending it 10 days to scale to the right. The same is done for activities 2 and 3 which both start at time zero. From the activity sheet it can be seen that activity 4 can only start when activities 1 and 2 are complete. Due to activity 1 having the longer duration, a vertical time line must be erected at the end of activity 1 and extended to meet the end of activity 2 produced by dotted line (the dotted line being the amount of float on activity 2). The time line prevents activity 4 starting before activity 1 is completed. Activity 4 can now be drawn. Activity 5 can be drawn straight after activity 4 as activity 5 is only dependent upon activity 4. Activity 6 is dependent upon activity 5 therefore straight line continues. Activity 7 is dependent upon the completion of activity 5 therefore a time line must be introduced to show the relationship between activities 5 and 7. From hereon each activity is considered in a similar fashion.

2.11   EFFECTS OF VARIATIONS AND DELAYS

2.11.1 Settling variations

The recognised procedure for settling variations in accordance with the R.I.B.A. Form of Contract clause 11(4) does not take all factors into account. Certain variations have a more far-reaching affect than those taken into account by adjusting the rate for measured work in the contract bills. When, for example, additional work is carried out, this will sometimes have the effect of lengthening the contract period. In these circumstances the contractor is entitled to reimbursement for the extra costs involved. Even when variations of omission are issued, it may not be a simple matter of deducting the prices of the respective items of measured work in the bill because the omission may disrupt the continuity of work for certain labour and plant, e.g. a gang of men or an item of plant may have a carefully arranged programme of work and the omission may result in men and plant being idle awaiting the next activity. In this case the labour and plant content of the omission should still be paid for by the client.

Clause 11 (6) makes provision for the architect to allow payment to the contractor in these circumstances. It will of course be necessary for the contractor to prove his case and to do this a programme is extremely useful.

2.11.2 Other factors causing delays

Many other factors can cause delays to the contractor besides variations, for which he does not, in many cases, receive full payment due to his inability to substantiate his claim, even though provision is made for payment under clause 24 of the R.I.B.A. form of contract. Examples of these factors are: −

> Late arrival of drawings, details and levels even when prior notice of these requirements has been given.
> Opening up work for inspection or testing.
> Discrepancies between drawings and contract bills.
> Delays caused by persons employed direct by the client etc.

The contractor can claim payments for losses caused by these factors where his progress has been disturbed.

2.11.3 Delays where no payment is warranted

Some delays do not of course warrant extra payment to the contractor, but it is nevertheless useful to have a means available by which the amount of extension of time can be justified. Typical delays in this category are exceptionally inclement weather, fire, strikes, shortage of materials or labour (if clause 23 (j) is applicable) which could not have been forseen. Extension of time can be granted for these and other reasons under clause 23 of the R.I.B.A. form of contract.

2.11.4 Assessing the effects of delays

It is often difficult to assess realistically the effect of delays on costs to the contractor, particularly when extension of time is involved. Without a programme it is virtually impossible. When a programme has been prepared and updated the problem of price adjustment, where appropriate, is made much easier. This particularly applies when the critical path method has been used. If the critical path programme is a contract document, it will almost certainly be used as a basis for calculating these adjustments. Even when the programme is not a contract document, providing the contractor updates it, he can present it to the quantity surveyor or architect as appropriate, whenever a variation order has the affect of interrupting the programme and it will prove to be formidable evidence in support of a claim.

On a critical path programme the effect can be clearly seen and variation orders or other causes which affect activities on the critical path will often warrant an extension of time and, possibly, subsequent additions to the contract sum.

2.11.5 Settling a variation

Figure 2.14, which represents the critical path programme for the village hall project will be used to illustrate the effects of a variation order on costs and contract period.

Assume a variation order is issued for additional screed (activity 16–18) and floor finish (activity 19–20) under the stage area. (This extra flooring is considered necessary because the client will require the stage to be moved on some occasions and wishes therefore to have the whole floor covered.)

The additional screed and floor finish will increase the total amount in the bill by one fifth and direct costs to the contractor will also increase by one fifth. These costs will be covered by adjusting the price included in the contract bills which will also increase by one fifth. However, both these activities are on the critical path and the time required for each will also increase by one fifth. This will result in an extension of time of $(1/5 \times 4) + (1/5 \times 12) = 3\frac{1}{5}$ days. The contractor may therefore require this extension of time plus the extra cost of overheads included in the preliminaries and related to the contract period.

It may of course be possible to avoid extending the contract period by over-time working or bringing in extra men. This will however increase costs and if the contractor agrees to take this action he is entitled to payment for the extra costs. This would probably be a matter for negotiation between the contractor and the quantity surveyor.

The example given is of course a very simple one, and in practice the problem may be more difficult. However the above procedure can be followed and differences of opinion should be much easier to resolve.

2.11.6. Assessing costs using bar charts

Bar charts can be used in a similar manner to that shown above but it may be more difficult to 'prove' that a particular activity is on the critical path and will therefore affect the project duration. A good updated bar chart is however far superior to no programme at all.

3: Work Study

3.1. INTRODUCTION

The human aspect is a most important part of work study, both on site and at higher management level, and the subject consists just as much of human relations (i.e. problems of communication and eliciting co-operation) as it does of techniques. In this chapter only the application of work study techniques will be considered; for a more detailed consideration of the subject as a whole, other texts listed in the bibliography should be consulted.

3.2 PURPOSE OF WORK STUDY

The purpose of work study is the provision of factual data to assist management in making decisions and to enable them to utilize with the maximum of efficiency all available resources (i.e. labour, plant, materials and management) by applying a systematic approach to problems instead of using intuitive guesswork.

3.3 BREAKDOWN OF WORK STUDY

Work study has two main aspects, method study and work measurement, which are very closely related. For convenience they will be considered separately in this text, but their interdependence must be appreciated at all times.

3.4 METHOD STUDY

The aim of method study is to provide information which will assist management in taking decisions related to the method it is proposed to use by making a systematic analysis of a problem and developing alternative methods, thus determining the optimum layouts and the most effective use of resources.

There are six basic steps involved in carrying out a method study:

1. Select the work to be studied and define the problem.
2. Record the relevant facts using the recording techniques.
3. Examine these facts critically and without bias to ascertain whether each element in the work is necessary, asking a series of questions and examining alternative solutions.
4. Develop the best method from the alternatives and submit this to management for approval.
5. Install the new method. Installation follows a decision by management to accept the method.
6. Maintain the new method. This consists of checking that the new method is adhered to by regular inspection on site or by watching output records which will indicate deviations from the standards set.

3.4.1 Recording techniques

There are three main groups of recording techniques available for setting down a problem. These are:

1. Charts	Outline process charts
	Flow process charts (man-type, material-type and
	equipment-type)
	Two-handed process charts
	Multiple-activity charts
2. Diagrams and models	Flow diagrams
(Two- and	String diagrams
three-dimensional)	Cut-out templates (two-dimensional models)
	Models (three-dimensional)
3. Photographic	Photographs
	Films

Templates, i.e. silhouettes of plant, equipment, storage areas etc., are very useful when considering layouts, and it is also possible to obtain charting kits, which consist of translucent sheets and stick-on templates that may be used for die-line prints. This is an excellent method of recording when it is desired to compare one method with another.

3.4.1.1 Flow process charts
This technique can be useful in helping to solve problems of layout such as those in site workshop areas where the operations are likely to be repetitive.

All activities are shown by means of symbols with a description against each. This is a very simple method of showing a sequence of work, and employs the following symbols:

◯   Operation   ▽   Storage e.g. stored for future use

▢   Inspection   D   Temporary storage or delay

▷   Transport

The movements of men, materials or equipment are followed through a process and the symbols are used to indicate what is happening at the various stages. The distance travelled may be shown on the transport activities and the time for each element can be given.

3.4.1.2 Flow diagrams
A flow diagram can be used in conjunction with a flow process chart to show where the activities take place. The same symbols that are employed on flow process charts are used but in this case they are superimposed onto a drawing and the descriptions are not necessary. All movements and distances can be clearly shown.

3.4.1.3 String diagrams
This technique is very useful in solving problems of movement. It is applied to repetitive situations and is therefore most useful in working areas such as factories producing industrialized components, machine shops, pre-casting yards, steel bending areas on site etc. The diagram will show up points of congestion and any excessive distances travelled.

The procedure for improving a layout is to first draw to scale a plan of the area under consideration with the work places or stacking areas etc. and all changes in direction denoted by pins. String or thread is then tied round the starting point and passed from pin to pin showing movement. Men, materials or machines can be denoted by different coloured string. If the string is then measured, any excessive distances travelled will be obvious and any points of congestion will be seen on the diagram.

Routes which are travelled regularly should be kept as short as possible, and alternative methods can be examined to obviate the faults of the first (templates can be very useful here).

3.4.2 Critical examination.

The step examine is the key step in a method study, and consists of a detailed examination of every aspect of the work.

The purpose is to:

1. Establish the true facts surrounding the problem.

2. Establish the reasons for these facts and determine whether they are valid.

3. On this foundation to consider all the possible alternatives and hence the optimum solution.

Examination is carried out by a questioning technique for which a critical examination sheet is very useful (Fig. 3.1). There should be liaison with the site management personnel who will carry out the work and they should be encouraged to put their ideas forward.

3.4.3 Site layout problems

When deciding on the relative positions of plant, working areas, storage areas etc. on site, reference has to be made to the various activities to be performed, i.e. the number and types of activity and the minimum transports and storages required. This basic information can be provided from materials flow process charts, which may be used in conjunction with a scale drawing of the site layout and paper templates. String diagrams can then be used either to evaluate or to visually represent the movement intensity for the alternative arrangements of the templates. These methods were used in the selection of the site layout in chapter 1 (Fig. 1.27).

3.4.4 Steel cutting and bending area

Flow charts, flow diagrams and string diagrams are used to assist in considering the layout of the steel bending area in the site for the amenity centre and office block (Fig. 1.27).

The space required for storing steel in the form it is delivered and after it has been prepared will be determined by the amount of steel involved and the timing of deliveries. A power bender will be required, but a hand cutter should be adequate for cutting purposes as large diameter bars and bars for the floor slab are being purchased cut to length.

The site manager has suggested the layout shown in Fig. 3.2 and a material-type flow diagram has been superimposed on it; a flow process chart has also been drawn up for this arrangement (Fig. 3.3).

The chart and diagram are now critically examined to find improvements, and from the flow diagram (Fig. 3.2) it can be seen that a fair amount of movement is involved in transporting steel to and from the various positions where operations would take place. A string diagram is therefore produced to show the movement of materials for one day (Fig. 3.4) and it becomes evident that many long distances are travelled an excessive number of times. Alternative layouts are examined to eliminate long transports on frequently used routes, and Fig. 3.5 shows the string diagram for the selected layout, with shortening of the routes travelled most frequently; the flow diagram for this layout is shown in Fig. 3.6. If required, different coloured string can be used to show the movement of bars cut to length on delivery and those prepared on site etc.

CRITICAL EXAMINATION SHEET

DESCRIPTION OF OPERATION:
DESCRIPTION OF ELEMENT:

INVESTIGN	PRIMARY QUESTION	SECONDARY QUESTION	POSSIBLE ALTERNATIVES	SELECT ALTERNATIVES
purpose	what is acheived	is it necessary if so why	what else could be done	what should be done
place	where is it done	why there	where else could it be done	where should it be done
sequence	when is it done	why then	when else could it be done	when should it be done
person	who does it	why him	who else could do it	who should do it
means	how is it done	why that way	how else could it be done	how should it be done

Fig. 3.1

Fig. 3.2 Flow diagram
Cutting and bending area

Had the layout suggested by the site manager not been put forward, the first stage would have been to produce an outline process chart, followed by flow process charts, flow diagrams, and string diagrams.

3.4.5 Multiple activity charts

This technique is used to help solve problems when a number of subjects (operatives, machines or equipment) are dependent on each other. It shows the occupied time (divided into elements of work if necessary) and unoccupied time for the subjects in both present and proposed methods. The subjects are recorded on a common time scale and periods of unoccupied time can be readily seen; alternatives may then be considered, the aim being to balance the work content for the subjects. It is necessary to plot more than one cycle of work as the first taken may not be representative.

CHART BEGINS: steel delived by lorry
CHART ENDS: steel awaiting collection
by crane

DATE:

CHARTED BY:

random & cut to length
steel delivered

unloaded & sorted

cut to length &
unbent steel
(for floor slabs)

random lengths
(for beam & columns)

to storage racks
for finished steel

to storage racks

stored

stored

to hand cutter

cut to length

to temporary store

awaiting bending

to bending machine

bent

cut & bent
steel (for beams)

cut & bent steel
(for columns)

to storage racks
for finished steel

to temporary
store

stored

awaiting
fabrication

to fabricating
area

fabricate
columns

to storage
area

stored

SUMMARY

○   4

□   1

▷   10

▽   4

D   2

Fig. 3.3 Flow process chart
Material type
Cutting and bending area

Fig. 3.4 String diagram
Cutting and bending area

3.4.5.1 Laying the concrete upper floor slabs of amenity centre and office block
Multiple activity charts are to be used to help increase efficiency in laying the
upper concrete floor slabs in the amenity centre and office block. The operation
is at present carried out by two men at the mixer (one driver and one loader),
the tower crane transporting the concrete to the required position, and one
labourer discharging the concrete from the skip and spreading and levelling it.
It is assumed that a time study has been carried out and gives the information
shown below (the times stated are hypothetical).

Man/machine and element	Code	Description	Time in standard minutes
Loader			
Load hopper	L	Load mixer hopper with fine and coarse aggregate using hand scraper	2.00

Mixer driver
    Add cement   ct   Add cement to fine and coarse   0.25
                                         aggregate in hopper (from silo
                                         above mixer)

Mixer driver
    Raise hopper   R   Raise hopper up to opening in   0.25
                                          mixer drum

Mixer driver and mixer
    Mix concrete   M $\left\{\begin{array}{l} \\ T \\ \\ \end{array}\right.$ $\left\{\begin{array}{l}\text{Tip contents into drum} \\ \text{Add water} \\ \text{Mix concrete}\end{array}\right.$ $\left.\begin{array}{r}0.20 \\ 0.05 \\ 1.75\end{array}\right\}2.00$

Mixer driver
    Lower hopper   H   Lower hopper from opening in   0.25
                                          mixer drum down to ground level

Mixer driver
    Discharge concrete   D   Position crane skip and discharge   0.50
                                          contents of mixer drum into it

Crane and driver
    Deliver concrete   DC   Lift concrete in skip, slew and   1.50
                                          deliver concrete to placing position

Spreader
    Discharge concrete   DL   Discharge concrete from crane skip   0.25

Spreader
    Spread concrete   SC   Spread and vibrate concrete   5.50
                                          between reinforcing steel and tamp

Crane and driver
    Return empty skip   RE   Return empty skip from placing   1.50
                                          position to concrete mixer

3.4.5.2 Method 1

The multiple activity chart (Fig. 3.7) shows the operation as carried out at present. One man loads the hopper on the mixer with sand and coarse aggregate. The mixer driver checks the weight on a dial and discharges the cement from the silo above the hopper; he next elevates the hopper and discharges the contents into the mixer drum, feeds water into the drum, and lowers the hopper. When the concrete is mixed, the mixer driver places the skip in position and then discharges the concrete from the mixer into it; the crane then lifts the concrete in the skip and transports it to the desired position. Finally a spreader discharges the skip and spreads the concrete, the crane returning with empty skip to the mixer.

    The chart shows that the time cycle of 5.75 standard minutes is determined by the spreader, and this results in unoccupied time for other operatives and machines.

3.4.5.3 Method 2

The multiple activity chart (Fig. 3.8) shows an alternative method. Two spreaders are used here to discharge and spread concrete from the crane skip, so that the crane now determines the time cycle of 3.75 standard minutes — a marked increase in efficiency.

Fig. 3.5 String diagram
Cutting and bending area

3.4.5.4 Method 3
The multiple activity chart (Fig. 3.9) shows another alternative. This procedure is similar to method 2 but the mixer driver also does the loading. The cycle time remains the same, and this results in a further increase in efficiency.

3.4.5.5 Other methods
Additional alternatives could be tried and one possibility is to use two crane skips, which might cut down some of the time at present wasted whilst the crane waits for the skip to be filled at the mixer. There would of course be extra time involved in attaching and detaching the skip, etc.

Fig. 3.6 Flow diagram
Cutting and bending area

Fig. 3.7

Fig. 3.8

MULTIPLE ACTIVITY CHART

OPERATION mixing, delivering & spreading concrete

PLANT 300/200 mixer, power scoop & cement silo, 500 kg. tower crane

GANG mixer driver, loader, tower crane driver, 2 spreaders

METHOD 3 1 at mixer, 2 spreading

time in standard minutes

0 1 2 3 4 5 6 7 8 9 10 11 12 13 14 15 16 17 18 19 20 21 22 23 24 25 26

MIXER — M ... DR M ... M M M M M M M DR M ... D

MXR DR & LOADER — L ... ct RTH ... L ... Lct RTH ... D L RTH ... D L RTH ... L RTH ... L RTH ... L RTH ... L ct

CRANE & DRIVER — D DC RE D DC RE D DC RE D DC RE D DC DL RE

cycle time 3·75 sms

SPREADER 1 — SC DL³ SC DL³ SC DL³ SC DL³ SC DL³ SC

SPREADER 2 — SC SC SC SC SC SC

CODE see text

UTILISATION cycle time 3·75 sms

mixer $\dfrac{2\cdot75 \times 100}{3\cdot75} = 73\%$

mxr dr & loader $\dfrac{3\cdot5 \times 100}{3\cdot75} = 93\%$

crane & driver $\dfrac{3\cdot75 \times 100}{3\cdot75} = 100\%$

spreader 1 $\dfrac{3 \times 100}{3\cdot75}$ 80%

spreader 2 $\dfrac{2\cdot75 \times 100}{3\cdot75}$ 73%

Fig. 3.9

3.5 WORK MEASUREMENT

The aim of work measurement is to determine the time it takes for a qualified worker to carry out a specific job at a defined level of performance and to eliminate ineffective elements of work. It seeks to provide the standard times for jobs and thus supplies basic, essential data for management.

3.5.1 Measurement techniques common in the construction industry

The work measurement techniques common in the construction industry are:
1. Time study
2. Activity sampling
3. Synthesis from standard and synthetic data
4. Analytical estimating.

3.5.2 Time study

The stages involved in carrying out a time study are:
1. Selecting the work to be measured
2. Analysing and breaking the work down into elements
3. Rating and timing each element
4. Extending the observed time to basic time
5. Selecting basic times, allocating allowances and building up the final standard.

Timing and rating are obviously carried out in the field and consequently in the example given it is assumed that the times and ratings have already been obtained.

3.5.2.1 Methods of timing

The methods of timing commonly used in the construction industry are cumulative timing and flyback timing.

Cumulative timing is the more common as it is better for observing a number of operatives in a gang and requires only an accurate wrist watch. The cumulative time is recorded after each element.

Flyback timing is carried out with a flyback stopwatch, the observer recording the time for each element as work proceeds. The watch has a flyback button on it which returns the hands to zero when pressed; on releasing the button the watch re-commences timing.

To check the accuracy, the start and finish times are taken and the difference between them is compared with the total of the readings.

3.5.2.2 Working up the standard time for an operation

To determine a standard time, a number of time studies of the operation will have to be completed; it is assumed that this has been done using cumulative timing.

In the example the following terms are used:

Element	A distinct part of an operation which can be easily defined.
Rating	The method used to take into account variations in working pace.
Cumulative time	The time recorded on the study sheet.
Observed time	The time taken to perform an element of work.
Standard rating	The average rate at which a qualified man will work given sufficient motivation and instruction.

Basic time   The time required for carrying out an element of work at standard rating.

Relaxation allowance   The time given to allow a worker to recover from the effects of fatigue and to attend to personal needs.

Contingency allowance   The time given to allow for occurrences throughout the day, e.g. a painter cleaning brushes.

3.5.2.3 Method used for calculating standard time

1. Calculate the observed time for each element by subtracting the previous cumulative time from the cumulative time for the element in question. Check that the total observed time equals the last cumulative time recorded (Fig. 3.10).

2. Extend the observed times to basic times, i.e. show all times at the standard rating by taking the observed rating into account. The formula for this calculation (Fig. 3.11) is

$$\frac{\text{Observed time} \times \text{observed rating}}{\text{Standard rating}}$$

3. Calculate the average of the basic times over the series of studies (Fig. 3.12).

4. Add the appropriate relaxation allowances and contingency allowances (Fig. 3.13), remembering that unoccupied time allowance and interference allowances must also sometimes be made.

This procedure will give the standard time for each element; to calculate the standard time for an operation, add the elements in the operation together.

NOTE: This example is very much simplified and the times are hypothetical.

Fig. 3.10

AMENITY CENTRE AND OFFICE BLOCK

OPERATION Concreting upper floor slabs

Element	Rating	Cumulative time	Observed time
Check time		2.75	2.75
A. Discharge crane skip	85	2.99	0.24
B. Spread concrete	75	4.89	1.90
	80		
	70		
C. Vibrate concrete	100	5.33	0.44
D. Level and tamp concrete	95		
	90		
	85		
	90	7.63	2.30
			7.63

Fig. 3.11

Element	Rating	Cumulative time	Observed time	Basic time
Check time	85	2.75	2.75	
A. Discharge crane skip	85	2.99	0.24	$\dfrac{0.24 \times 85}{100} = 0.20$
B. Spread concrete	75 80 70	4.89	1.90	$\dfrac{1.90 \times 75}{100} = 1.43$
C. Vibrate concrete	100	5.33	0.44	$\dfrac{0.44 \times 100}{100} = 0.44$
D. Level and tamp concrete	95 90 85 90	7.63	2.30 ——— 7.63 ——	$\dfrac{2.30 \times 90}{100} = 2.07$

Fig. 3.12

Element	Basic time	Total basic times	Frequency	Average
A. Discharge crane skip	0.20, 0.18, 0.19, 0.18, 0.21, 0.20, 0.18, 0.19, 0.19, 0.18	1.90	10	0.19
B. Spread concrete	1.43,.1.27, 1.31, 1.49, 1.45, 1.45, 1.72, 1.48, 1.80, 1.60	15.00	10	1.50
C. Vibrate concrete	0.44, 0.46, 0.44, 0.42, 0.44, 0.45, 0.45, 0.44, 0.43, 0.43	4.40	10	0.44
D. Level and tamp concrete	2.07, 2.13, 2.35, 2.35, 2.23, 2.27, 2.24, 2.00, 2.16, 2.20	22.00	10	2.20

Assume that the total relaxation allowance and contingency allowance are as follows:

Fig. 3.13

Element	Basic time	Total relaxation allowance	Contingency allowance	Total allowances	Standard time (minutes)
A. Discharge crane skip	0.19	25%	5%	30%	$\dfrac{0.19 \times 130}{100} = 0.25$
B. Spread concrete	1.50	30%	5%	35%	$\dfrac{1.50 \times 135}{100} = 1.95$
C. Vibrate concrete	0.44	26%	5%	31%	$\dfrac{0.44 \times 131}{100} = 0.58$

Element	Basic time	Total relaxation allowance	Contingency allowance	Total allowances	Standard time (minutes)
D. Level and tamp concrete	2.20	30%	5%	35%	$\dfrac{2.20 \times 135}{100} = 2.97$

The standard time for the operation is 0.25 + 1.95 + 0.58 + 2.97 = 5.75 standard minutes.

3.6   ACTIVITY SAMPLING

Activity sampling is also commonly known as snap observation studies or random observation studies. It is a very useful technique in the construction industry, and is used to determine the activity levels of machines and operatives. Using this method a number of subjects can be observed concurrently.

3.6.1 Basis of the method

In this procedure the principles of the statistical method of random samples are used. Snap readings are taken at intervals and the percentage of readings taken per element will give a result very near to the actual percentage on each providing the sample is big enough. From this it is possible to calculate the time taken on each element if necessary.

One observer can record the activities of a whole gang and readings can be almost continuous. The timing may be by either the fixed interval method (easier to use on site), or the random interval method, which is better for work of a cyclic nature.

3.6.2 Uses of activity sampling

On site activity sampling can be used:

1. To assess unoccupied time as a basis for analysing cause.

2. To find the percentage of time spent on each element of work in an operation by each operative and/or machine.

3. To find the percentage utilization of machines or operatives as a basis for cutting down unoccupied time (rated activity sampling can be used for this, which is an extention of the method).

3.6.3 An example on activity sampling

The following example (Fig. 3.14) shows the application of activity sampling to determine the proportions of productive and non-productive time for three excavators owned by a contractor.

A total of 999 observations were made on the three excavators and at each observation the machines were recorded as being productive, suffering from major delays, or suffering from minor delays.

Weather delays, opening up working areas, and machine repairs were defined as a major delay, whilst insufficient lorries, spotting lorries, trimming and clearing up, moving the machine, routine maintenance, and operator delays were held to constitute minor delays.

If it is felt that the delays are excessive, a further analysis can be carried out to ascertain the actual cause of delay under the more detailed headings listed above. As stated previously, the accuracy of the result is determined to some extent by the size of the sample and formulae are available which will give:

(a)   the number of observations to produce a given accuracy, and
(b)   the limit of accuracy obtained from the observations made

These formulae are:

$$\text{number of observations required} = \frac{4P\,(100 - P)}{L^2}$$

where P is the expected percentage activity and L is the limit of error (percentage ±)

$$\text{limit of accuracy} = 2\sqrt{\frac{P\,(100 - P)}{N}}$$

where N is the number of observations and P is the percentage of observations recorded for a particular activity, and ±L are the 95 per cent confidence limits.

These formulae will now be applied to the example to ascertain:

(a)   the number of observations necessary to give a limit of accuracy of ±3 per cent on minor delays, and

(b)   the accuracy achieved with the number of observations taken.

Before starting the calculations it is necessary to estimate what the percentage on the activity will be, either from a short preliminary study or from other information. In this example, an estimate of 33 per cent has been made.

(a)   Number of observations required

$$= \frac{4P\,(100 - P)}{L^2}$$

$$= \frac{4 \times 33\,(100 - 33)}{3^2}$$

$$= \frac{4 \times 33 \times 67}{9}$$

$$= 983 \text{ observations}$$

999 observations were actually taken

(b)   Limit of accuracy of result obtained (Fig. 3.14)

$$= 2\sqrt{\frac{P\,(100 - P)}{N}}$$

$$= 2\sqrt{\frac{21 \times 79}{999}}$$

$$= 2\sqrt{1.6}$$

$$= 2 \times 1.26$$

$$= \pm 2.52 \text{ per cent}$$

The formulae can also be applied to major delays and productive time if required.

Fig. 3.14

Number of tours of site	Excavator 1	Excavator 2	Excavator 3
1	P	SD	P
2	SD	P	LD
3	LD	LD	P
etc	etc	etc	etc
333	P	SD	LD

					Totals
Totals for	P	133	137	140	410
999	SD	73	70	67	210
Observations	LD	127	126	126	379

Productive time = P   Minor delays = SD   Major delays = LD

Calculations:

$$\text{productive time} \quad \frac{410 \times 100}{999} = 40 \text{ per cent}$$

$$\text{minor delays} \quad \frac{210 \times 100}{999} = 21 \text{ per cent}$$

$$\text{major delays} \quad \frac{379 \times 100}{999} = 38 \text{ per cent}$$

3.7   SYNTHESIS AND SYNTHETIC DATA

The purpose of synthetic data is to enable time values to be compiled for jobs where direct measurement is unnecessary or impracticable. Many operations are very similar to ones which have been carried out before, perhaps with variation due to physical dimensions etc. (examples are: sawing different lengths of timber and paving different areas); these various jobs have many common elements and a suitably referenced stock of data relating to these elements can be built up over a period of time and used for quickly compiling the work content of various jobs.

3.8   ANALYTICAL ESTIMATING

To calculate the work content of non-repetitive jobs, the work value can be compiled by analytical estimating, using whatever information is available from past time studies or standard data and estimating the time for the remaining elements.

3.9 FINANCIAL INCENTIVES

3.9.1 Aims of financial incentive schemes

The aim of a sound financial incentive scheme is to increase productivity by increasing the efficiency of the individual, and enable him to increase his earnings.

A good scheme will achieve these objectives, but a poor one will cause nothing but trouble. Incentive schemes should be very closely related to short term planning and cost control. Careful supervision is necessary to ensure quality is maintained.

3.9.2 Features of a sound scheme

The main features of a sound financial incentive scheme are:

1. The amount of bonus paid to the operatives should be a direct proportion of the time saved and there should be no limit to the amount that can be earned.

2. Targets should be issued for all operations before work commences. The extent and nature of the operation should also be clearly understood. The length of time for an operation should ideally be kept to about two days.

3. Targets should be set for small gangs (there are exceptions to this in repetitive work similar to flow production when group bonuses may be more appropriate).

4. Targets should not be altered during an operation without agreement of both parties.

5. The system used for calculating bonus earnings should be easily understandable by all persons concerned.

6. Operatives must not be penalised for lost time which is caused by delays outside their control.

7. Bonus should be paid weekly.

3.9.3 Basis for setting targets

3.9.3.1 The estimate

For large projects where the estimate is based on a bill of quantities the rates used are not really suitable for direct use as bonus targets. The figures used in the bill are average, e.g. brickwork operations will vary in difficulty and, therefore, speed of completion depending on their situation. Brickwork in small panels takes longer per unit area to build, than a large expanse of uninterrupted brickwork. This is often not clear from the traditional Bill of Quantities. In addition it is a laborious job to collect bill items together which are relevant to a particular operation.

For small projects where the builder has to prepare the estimate from drawings it is easier to extract the labour content of operations and set targets based on this. By this means a check can easily be made on profits. The labour content in the estimate should of course be based on data from past projects.

3.9.3.2 Data from past projects

This is probably the most reliable information reasonably available to most firms where work measurement has not yet been used. The data will be built up over a period of years and can be used as a basis for estimating and as a basis for setting bonus targets, etc. Great care is necessary in analysing this data as it must be appreciated that the level of productivity, i.e. *the rating* will vary considerably. The conditions under which operations are carried out should also be analysed.

3.9.3.3 *Work measurement*

This is by far the best method but as stated above, is not at present available to the vast majority of firms. The targets based on work measurement are more accurate and the method to be used is taken into account. Targets based on work measurement are less likely to be questioned by the operatives, but where targets are questioned a check should be made before making adjustments.

3.9.3.4 *Other methods*

Experience, bargaining and other methods are used but are not very reliable generally and will not be considered here.

3.9.4 Expected earnings when using bonus schemes

The general level of earnings may well be fixed by competition from local firms and this may determine the basis for setting targets.

It is generally felt, however, that an operative should be paid 1/3 of his basic weekly earnings, as a bonus when he works at a rating of 100, and the following discussion is based on this.

3.9.5. Rate of payment for time saved

Operatives are paid differing rates of payment for time saved depending on circumstances. This will affect the targets set in order to allow the 1/3 bonus at 100 rating.

Examples of rates of payment which could be used are: —

1. Payment of full rate for each hour saved (100% system)
2. Payment of 75% of hourly rate for each hour saved (75% system)
3. Payment of 50% of hourly rate for each hour saved (50% system)
4. Payment of 25% of hourly rate for each hour saved (25% system)

When the payments made for time saved are less than the full hourly rate the systems are known as *Geared Systems.*

3.9.6 Effects of different rates of payment on earnings and costs (see Fig. 3.15).

3.9.6.1 *Effects on earnings*

Assume an operation takes x hours to perform at 100 rating and at this $\frac{1}{3}$ bonus will be earned.

1. *Payment of all time saved (100% system).*
$$\text{Target} = x + \frac{x}{3} = 1\tfrac{1}{3}x$$

2. *Payment of 75% of time saved (75% system)*
$$\text{Target} = x + \frac{\tfrac{4}{3}}{3}x = 1\tfrac{4}{9}x.$$

3. *Payment of 50% of time saved (50% system)*
$$\text{Target} = x + \tfrac{2}{3}x = 1\tfrac{2}{3}x.$$

4. *Payment of 25% of time saved (25% system)*
$$\text{Target} = x + \tfrac{4}{3}x = 2\tfrac{1}{3}x$$

Rating at which bonus will start to be earned
100% system:

target is $\frac{4}{3}$ of operations time at 100 rating

∴ bonus starts at $\frac{3}{4} \times 100 = 75$ rating.

75% system:

   target is $\frac{13}{9}$ of operations time at 100 rating

   ∴ bonus starts at $\frac{9}{13}$ × 100 = 69.2 rating

50% system:

   target is $\frac{5}{3}$ of operations time at 100 rating

   ∴ bonus starts at $\frac{3}{5}$ × 100 = 60 rating

25% system:

   target is $\frac{7}{3}$ of operations time at 100 rating

   ∴ bonus starts at $\frac{3}{7}$ × 100 = 42.9 rating

3.9.6.2 *Effects on costs*

It is assumed that the labour content of estimated prices are based on outputs which approximate to a rating of 75. The following comparisons show the effects of different systems of payment on actual direct costs.

1. *100% system*

   At $37\frac{1}{2}$ rating, cost

   $$=\frac{75}{37\frac{1}{2}} \times 100 = 200\%$$

   At 40 rating, cost

   $$=\frac{75}{40} \times 100 = 187\frac{1}{2}\%$$

   At 60 rating, cost

   $$=\frac{75}{60} \times 100 = 125\%$$

   At 75 rating and above cost is constant at 100%.

2. *75% system* — target $1\frac{4}{9}x$

   Cost as above until bonus is earned at 42.9 rating.

   At 75 rating — time required = $\frac{4}{3}x$

      time saved = $1\frac{4}{9}x - \frac{4}{3}x = \frac{1}{9}x$

      bonus = $\frac{3}{4} \times \frac{1}{9}x = \frac{1}{12}x$

      cost $=\dfrac{\frac{4}{3}x + \frac{1}{12}x}{\frac{4}{3}x} = 106\frac{1}{4}\%$

   At 100 rating — time required = x

      time saved = $1\frac{4}{9}x - x = \frac{4}{9}x$

      bonus = $\frac{1}{3}x$

      cost $=\dfrac{x + \frac{1}{3}x}{\frac{4}{3}x} = 100\%$

At 125 rating – time required $= \frac{4}{5}x$

 time saved $= 1\frac{4}{9}x - \frac{4}{5}x = \frac{29}{45}x$

 bonus $= \frac{29}{60}x$

 cost $= \dfrac{\frac{4}{5}x + \frac{29}{60}x}{\frac{4}{3}x} = 96\frac{1}{4}\%$

At 150 rating – time required $= \frac{2}{3}x$

 time saved $= 1\frac{4}{9}\dot{x} - \frac{2}{3}x = \frac{7}{9}x$

 bonus $= \frac{7}{12}x$

 cost $= \dfrac{\frac{2}{3}x + \frac{7}{12}x}{\frac{4}{3}x} = 93\frac{3}{4}\%$

At 175 rating – time required $= \frac{4}{7}x$

 time saved $= 1\frac{4}{9}x - \frac{4}{7}x = \frac{55}{63}x$

 bonus $= \frac{55}{84}x$

 cost $= \dfrac{\frac{4}{7}x + \frac{55}{84}}{\frac{4}{3}x}x = 92\%$

3. *50% system* – target $= 1\frac{2}{3}x$

 Cost as above until bonus is earned at 60 rating.

 At 75 rating

  time required $= \frac{100}{75}x = 1\frac{1}{3}x$

  target $= 1\frac{2}{3}x$

  time saved $= 1\frac{2}{3}x - 1\frac{1}{3}x = \frac{1}{3}x$

  bonus $= \frac{1}{6}x$

  cost $= \dfrac{1\frac{1}{3}x + \frac{1}{6}x}{1\frac{1}{3}x} = 112\frac{1}{2}\%$

 At 100 rating

  time required $= x$

  time saved $= 1\frac{2}{3}x - x = \frac{2}{3}x$

  bonus $= \frac{1}{3}x$

  cost $= \dfrac{x + \frac{1}{3}x}{1\frac{1}{3}x} = 100\%$

 At 125 rating

  time required $= \frac{100}{125}x = \frac{4}{5}x$

  timed saved $= 1\frac{2}{3}x - \frac{4}{5}x = \frac{13}{15}x$

  bonus $= \frac{13}{30}x$

  cost $= \dfrac{\frac{4}{5}x + \frac{13}{30}x}{1\frac{1}{3}x} = 92\frac{1}{2}\%$

At 150 rating

time required = $\frac{100}{150}x = \frac{2}{3}x$

time saved = $1\frac{2}{3}x - \frac{2}{3}x = x$

bonus = $\frac{x}{2}$

cost = $\dfrac{\frac{2}{3}x + \frac{x}{2}}{1\frac{1}{3}x} = 87\frac{1}{2}\%$

At 175 rating

time required = $\frac{100}{175}x = \frac{4}{7}x$

time saved = $1\frac{2}{3}x - \frac{4}{7}x = \frac{23}{21}x$

bonus = $\frac{23}{42}x$

cost = $\dfrac{\frac{4}{7}x + \frac{23}{42}x}{1\frac{1}{3}x} = 84\%$

4. *25% system* — target = $2\frac{1}{3}x$

Cost as above until bonus is earned at 42.9 rating

At 50 rating

time required = $\frac{100}{50}x = 2x$

time saved = $2\frac{1}{3}x - 2x = 1\frac{1}{3}x$

bonus = $\frac{1}{12}x$

cost = $\dfrac{2x + \frac{1}{12}x}{1\frac{1}{3}x} = 156\frac{1}{4}\%$

At 75 rating

time required = $\frac{100}{75}x = \frac{4}{3}x$

time saved = $2\frac{1}{3}x - 1\frac{1}{3}x = x$

bonus = $\frac{x}{4}$

cost = $\dfrac{\frac{4}{3}x + \frac{x}{4}}{1\frac{1}{3}x} = 118\frac{3}{4}\%$

At 100 rating

time required = x

time saved = $2\frac{1}{3}x - x = 1\frac{1}{3}x$

bonus = $\frac{1}{3}x$

cost = $\dfrac{x + \frac{1}{3}x}{1\frac{1}{3}x} = 100\%$

At 125 rating

time required = $\frac{100}{125}x = \frac{4}{5}x$

times saved = $2\frac{1}{3}x - \frac{4}{5}x = 1\frac{8}{15}x$

bonus = $\frac{23}{60}x$

cost = $\dfrac{\frac{4}{5}x + \frac{23}{60}x}{1\frac{1}{3}x} = 88\frac{3}{4}\%$

At 150 rating

time required $= \frac{100}{150}x = \frac{2}{3}x$

time saved $= 2\frac{1}{3}x - \frac{2}{3}x = 1\frac{2}{3}x$

bonus $= \frac{5}{12}x$

cost $= \frac{\frac{2}{3}x + \frac{5}{12}x}{1\frac{1}{3}x} = 81\frac{1}{4}\%$

At 175 rating

time required $= \frac{100}{175}x = \frac{4}{7}x$

time saved $= 2\frac{1}{3}x - \frac{4}{7}x = \frac{37}{21}x$

bonus $= \frac{37}{84}x$

cost $= \frac{\frac{4}{7}x + \frac{37}{84}x}{1\frac{1}{3}x} = 76\%$

3.9.7 General comments

3.9.7.1 *The 100% system*

The 100% system is suitable where targets set are very accurate, i.e. in factory conditions when work measurement is used as a basis for the targets.

As shown in the graph the cost is constant when the rating exceeds 75. The operatives earn a greater bonus when the rating exceeds 100 when this system is used.

For site work it is very difficult to set really accurate targets, particularly in view of the fact that most firms do not use work measurement as a basis. If tight targets are set, the operatives want the targets adjusting. If the targets are too easy, the operatives earn high wages. Therefore management tend to loose in both cases. When productivity is below 75 rating costs are high but when productivity is high, management do not make any gains to counteract the losses.

3.9.7.2 *Geared systems*

Costs are higher than the 100% system below 100 rating down to the level at which bonus earnings start. Costs break even at a rating of 100. Above 100 rating management makes a saving which balances some of the losses at lower productivity.

(a) *75% system*

This system is sometimes used in the Construction industry on repetitive construction where reasonably accurate targets can be set, e.g. rationalised traditional housing.

(b) *50% system*

This is the system most used in the construction industry.

(c) *25% system*

This system is not much used. There is little incentive for the operatives to work at high ratings when they only get 25% of the time saved. In addition the system is very costly at low ratings.

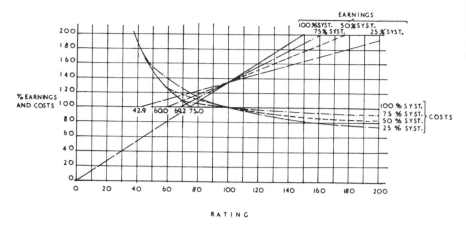

RATING

Fig. 3.15 (Earnings/cost chart)

4: Statistics

4.1 INTRODUCTION

The main purposes of statistics and statistical method are threefold:

1. To record past events. The statisticians and administrators are interested in the past and present as well as the future. By comparing the present with the past it may be possible to isolate trends and tendencies so that a forecast can be arrived at.

2. To identify what is happening in what appears to be an irregular and unrelated mass of information and present the data in such a form that it is possible to draw conclusions and take any required action.

3. To create methods of comparing data generally.

4.2 ANALYSIS AND PRESENTATION OF DATA

4.2.1 Analysis of data

Data can be analysed in many different ways, the method used depending upon the complexity of the information and the use to be made of the resulting analysis. The purpose of analysing data is to achieve a better understanding of it and to facilitate comparison with other data; it is usual practice to compare groups of data on the basis of their average, frequency distribution, and standard deviation.

4.2.1.1 Averages

Data can be analysed to find its average value and can be compared with the average of other data or with individual values in the group of data. There are three main types of average, the mean, the mode and the median, each of which has its usefulness in different fields of enquiry.

The mean is easy to calculate but becomes unduly affected by extreme values and consequently can be misleading. It is the arithmetical average — the sum of all values in a group divided by the number of values concerned.

The mode is unaffected by abnormal values in the group, and unlike the mean represents an actual value — the most frequently occuring one in a group.

The median is the middle value in a group and also eliminates the effect of extreme values. Half the observations are greater than it and half the observations are smaller; when the number of values is even, it is usual to take the mean of the two middle values to find the median.

4.2.1.2 Frequency distribution

Another important method of analysis is to sort data into predetermined groups having certain limits so that one group can be compared with another on the basis of frequency of occurrence in each group. Using this method of analysis it is also possible to compare one list of results with another by analysing each list into a number of groups: comparison is then carried out by considering the number of occurrences in each group in each list. For example, in the section on quality control 25 test cube results have been taken and sorted into groups having various predetermined strengths (Fig. 4.15.2), the analysis being represented by the use of a histogram. It would be possible to analyse another 25 test cube results in the same manner using the same group limits, and then to compare the first 25 results with the second 25 results on the basis of frequency distribution, i.e. the manner in which the results fall into the predetermined groups.

A complete example of this method of analysis is shown in the section on quality control (section 4.3.4).

4.2.1.3 Standard deviation

This is another useful way of comparing one set of data with another. Like the mean, mode and median, it expresses the characteristics of a group of data in one numerical expression (although with greater precision), and is used to measure the deviation of items in a set of data from the mean value of that set of data.

By using these measures of the central tendency, two sets of data could appear to be identical because they possess the same average, despite the fact that the range and distribution of results is very different. This contrast would be brought out by finding the standard deviation (Fig. 4.15.5) for each set of data and comparing them on this basis.

4.2.2 Presentation of data

The analyst may discover that he can achieve a clearer understanding of the meaning of the analysed data if he illustrates the results by some suitable graphical technique, a procedure which is often also useful for presenting the data to others who are less familiar with it.

Data must always be presented in a form that is easily understood by the recipient, and only data that is pertinent to the recipient's particular interest should be presented to him. For example, the managing director of a large construction firm would require to know weekly whether each of his projects was operating at a profit or a loss; information about variation in unit costs would consequently be unimportant to him, and it would be pointless to present him with a weekly cost control sheet for each project as he would only require information on total value and total cost from this sheet.

4.2.2.1 Tabulation (Fig. 4.1)

When using this form of presentation it must be borne in mind that however well presented they may be, columns of figures have very little meaning to some recipients. Tables should always be kept as simple as possible so that they do not become unintelligible due to the mass of data presented, and a useful addition in complex tables is the percentage (which has been used in Fig. 4.1 and makes it far easier to appreciate the proportions). Various types and weights of print can be used to attract attention to important features in a table, and a system of bold type for totals is shown in Fig. 4.1.

4.2.2.2 Pictograms

This form of presentation has a limited use, since it stimulates interest in a topic rather than presenting detailed facts. In Fig. 4.2.1 comparisons are made by counting the number of units shown for each group being compared. In Fig. 4.2.2 (top) the respective quantities are represented by the height of each unit, but in this form of presentation there is a danger of areas being compared and creating a false impression. This problem has been overcome in Fig. 4.2.2 (bottom) by standardizing on the width of each unit's base.

4.2.2.3 Pie chart (Fig. 4.3)

This is a very simple form of presentation that creates an immediate impression. These diagrams can vary from ones which have three or four slices to ones which have many slices but it must be remembered that a large number of slices make the chart more difficult to assimilate at a glance. This can be overcome to some extent by the use of colours or shading and the use of percentages. In the pie chart, each slice is proportional to the quantity represented.

Fig. 4.1

Year	All dwellings					Houses					Flats				
	1 bed	2 beds	3 beds	4 beds +	Total	1 bed	2 beds	3 beds	4 beds +	Total	1 bed	2 beds	3 beds	4 beds +	Total
1956	14864 10.62	48523 34.67	73876 52.78	2702 1.93	139965 100%	4606 4.49	28614 27.90	67016 65.34	2331 2.27	102567 100%	10258 27.43	19909 53.24	6860 18.34	371 0.99	37398 100%
1957	18287 13.29	49361 35.88	67017 48.71	2919 2.12	137584 100%	5557 5.84	27725 29.13	59319 62.33	2566 2.70	95167 100%	12730 30.01	21636 51.01	7698 18.15	353 0.83	42417 100%
1958	20132 17.79	40863 36.12	49681 43.91	2470 2.18	113146 100%	6121 8.39	21217 29.08	43415 59.50	2214 3.04	72967 100%	14011 34.87	19646 48.90	6266 15.60	256 0.64	40179 100%
1959	21840 21.96	34339 34.53	41347 41.57	1930 1.94	99456 100%	6832 11.56	16133 27.30	34445 58.29	1681 2.84	59091 100%	15008 37.18	18206 45.10	6902 17.10	249 0.62	40365 100%
1960	27057 26.21	35463 34.35	39050 37.83	1665 1.61	103235 100%	8941 15.03	16708 28.09	32323 54.34	1513 2.54	59485 100%	18116 41.41	18755 42.87	6727 15.38	152 0.35	43750 100%
1961	24257 26.17	29690 31.97	37018 39.86	1915 2.06	92880 100%	7885 14.69	13589 25.32	30755 57.31	1430 2.66	53659 100%	16372 41.74	16101 41.05	6263 15.97	485 1.24	39221 100%
1962	28971 27.51	34012 32.30	40275 38.25	2044 1.94	105302 100%	8460 14.57	14163 24.39	33707 58.05	1740 3.00	58070 100%	20511 43.43	19849 42.02	6568 13.91	304 0.64	47232 100%
1963	27079 27.91	31347 32.31	36539 37.66	2050 2.11	97015 100%	7041 13.92	12279 24.27	29434 58.17	1845 3.65	50599 100%	20038 43.17	19068 41.08	7105 15.31	205 0.44	46416 100%
1964	32952 27.58	41291 34.56	42740 35.77	2485 2.08	119468 100%	8385 13.83	15263 25.18	34825 57.44	2151 3.55	60624 100%	24567 41.75	26028 44.23	7915 13.45	334 0.57	58844 100%
1965	36351 27.33	47723 35.88	46226 34.75	2724 2.05	133024 100%	7474 11.80	16671 26.31	36847 58.16	2365 3.73	63357 100%	28877 41.45	31052 44.57	9379 13.46	359 0.52	69667 100%

NEW DWELLINGS COMPLETED FOR LOCAL AUTHORITIES

Analysis by bedroom type in England and Wales (based on the *Annual Digest of Statistics* [H.M.S.O., 1967])

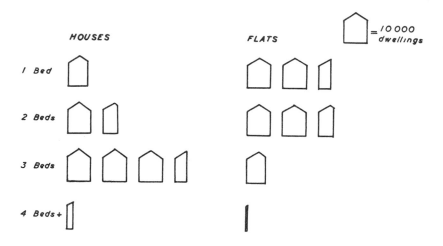

NEW DWELLINGS COMPLETED FOR LOCAL AUTHORITIES 1964
Analysis by Bedroom Type
England & Wales

Fig. 4.2.1 (based on Fig. 4.1)

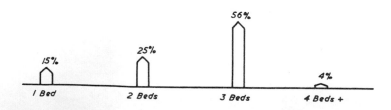

NEW HOUSES COMPLETED FOR LOCAL AUTHORITIES 1964
Analysis by Bedroom Type
England & Wales

Fig. 4.2.2 (based on Fig. 4.1)

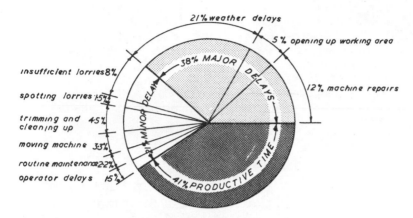

Fig. 4.3 Breakdown of productive and non-productive time for excavators owned by A.B.C. Builders Ltd.

4.2.2.4 Bar charts (Figs. 4.4.1, 4.4.2, and 4.4.3)

This technique is used to compare items that have a common characteristic (usually time). The principle is to compare items by area of bar, so for convenience it is better to standardize on width so that the bars can be compared by their length.

Bar charts take many forms with various horizontal or vertical arrangements, and compound bars are also frequently used. Fig. 4.4.1 (top) shows horizontal compound bars, whereas Fig. 4.4.1 (bottom) demonstrates the same information broken down into individual bars, and Fig. 4.4.2 gives the same information in one compound bar. To make the chart more impressive, the major divisions of the compound bar have been shaded differently, thus drawing attention to the major divisions.

Figure 4.4.3 shows information abstracted from Fig. 4.1. This figure again uses the compound bar chart to present the information, Fig. 4.4.3 (top) presenting the information on a numerical basis whilst Fig. 4.4.3 (bottom) presents the same information on a percentage basis. It is far easier to gain an impression of the relative proportions involved from the latter than from the former.

It is interesting to compare Figs. 2, 5, and 6 to ascertain which form of presentation creates the greatest impression, since these three figures present the same information in different ways.

4.2.2.5 Histogram

This is a type of vertical bar chart in which it is common to have the bars touching each other on either side. Here again comparison is by area, so standardization of width is important. The chart is used to show the relationship between two sets of quantities and is often used to present the frequency of certain occurrences and demonstrate frequency distributions.

One example of a histogram is the labour demand chart (Fig. 4.5.1), in which the planned labour force for the amenity centre is indicated. When the project is under way the actual labour force can be plotted on the same chart, thus indicating any discrepancies

It will be noted that a trough appears between weeks 20 and 25, and consideration should be given to amending the overall programme to avoid this, since its effect is to increase the proportion of direct to indirect wages. It may be found (as in this case) that the cause was the method of construction dictated by the design of the building, making alternative solutions too expensive: had the plastering and tiling been sub-contracted, which is the usual practice for many contractors, these troughs would not have appeared. In this type of chart the vertical lines marking the sides of the bars are usually omitted for clarity.

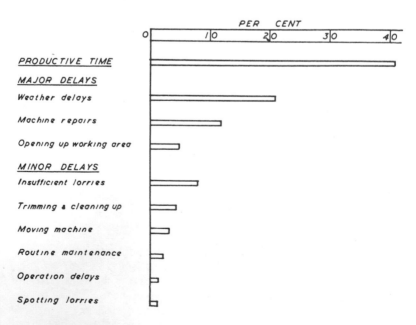

BREAKDOWN OF PRODUCTIVE & NON PRODUCTIVE TIME
FOR EXCAVATORS OWNED BY A.B.C. BUILDERS LTD.

Fig. 4.4.1 (based on Fig. 4.3)

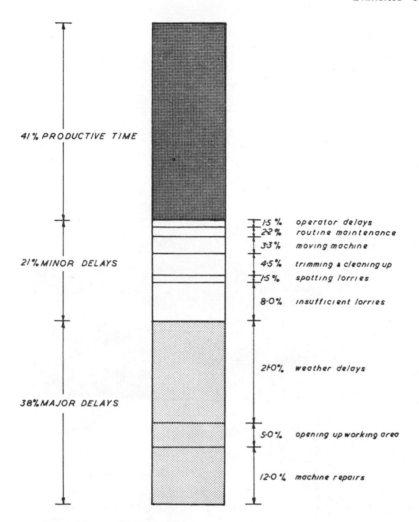

41% PRODUCTIVE TIME

21% MINOR DELAYS

38% MAJOR DELAYS

1·5 %   operator delays
2·2 %   routine maintenance
3·3 %   moving machine

4·5 %   trimming & cleaning up

1·5 %   spotting lorries

8·0 %   insufficient lorries

21·0%   weather delays

5·0 %   opening up working area

12·0 %   machine repairs

**BREAKDOWN OF PRODUCTIVE & NON PRODUCTIVE TIME
FOR EXCAVATORS OWNED BY ABC BUILDERS LTD.**

Fig. 4.4.2 (based on Fig. 4.3)

When a number of histograms are to be compared with each other, the three dimensional technique is useful (Fig. 4.5.2).

4.2.2.6 Array

This chart is similar to the bar chart but quantities are represented by the length of a line rather than the area of a bar. The chart is used to represent a large number of items which are arranged in order of magnitude, with the smallest item at the top (Fig. 4.6).

4.2.2.7 Graphs

A graph is a line which is plotted on a chart field. The chart field is generally divided into equal vertical divisions and equal horizontal divisions (an arithmetic chart field), but other chart fields are sometimes used which are divided into

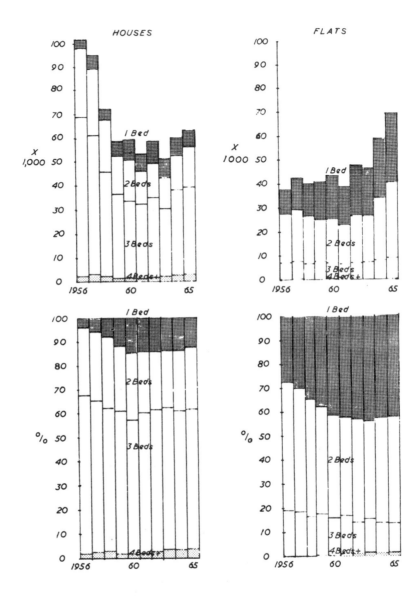

NEW DWELLINGS COMPLETED FOR LOCAL AUTHORITIES

Analysis by Bedroom Type

England & Wales.

Fig. 4.4.3 (based on Fig. 4.1)

Note

The histogram shows at a glance that a trough appears between weeks 20 and 25. Consideration should be given to amending the overall programme to avoid this trough. The effect of the trough is to increase the proportion of direct to indirect wages. It may be found as in this case that the cause was the method of construction dictated by the design of the building and alternative solutions could prove too expensive.

Had the plastering and tiling been sub-contracted, which is usual practice for many contractors, these troughs would not have appeared.

Fig. 4.5.1

equal horizontal divisions but have the vertical scale calibrated by logarithms (semi-log chart field).

The following general rules should be observed when constructing chart fields and drawing graphs.

1. The axes should be suitably annotated to indicate what they represent and have sufficient calibration to facilitate easy reading.

2. Care should be taken in selecting the scales used, and they should be clearly indicated on the axes.

3. Starting each scale at zero will avoid a false impression being imparted.

4. Correct use of the squares on the chart field will facilitate easier reading.

5. When two scales have to be used for the vertical axis, confusion is avoided if these scales can be kept on either side of the chart field (by way of illustration this has been done in Fig. 4.9 but not in Fig. 4.10).

6. When cost is one variable represented, this should be plotted on the vertical axis so that increases and decreases can easily be appreciated. When time is a variable this is usually plotted on the horizontal axis.

7. The figures from which the chart is made up should be clearly indicated on the chart by dots or crosses to ensure that values are not missed when the graph is being drawn.

8. When a number of graphs are being compared on the same chart field, advantage should be taken of the use of colours or different line codes. However, the number of lines used must be restricted when they fall in the same part of the chart field.

NEW FLATS COMPLETED FOR LOCAL AUTHORITIES

Analysis by bedroom type

England & Wales

1 Bed

2 Beds

3 Beds

Fig. 4.5.2 (based on Fig. 4.1)

Concrete erectors & assemblers	3·1
Glaziers	3·1
Crane drivers	5·8
Formwork erectors	6·6
Masons	6·9
Steelfixers	6·9
Floor & Wall tilers	7·9
Paviours	8·8
Steel erectors & sheeters	8·8
Roofing slater & tilers	11·2
Scaffolders	11·7
Plasterers	26·7
Heating & Ventilating engineers	29·5
Mechanical plant operators	36·4
Plumbers & Gass fitters	53·3
Electrician	66·2
Bricklayers	93·7
Painters	109·9
Carpenters & Joiners	164
Labourers	314

NUMBER EMPLOYED X.1,000

OPERATIVES EMPLOYED BY CONTRACTORS   SEPT 1966

N.B.   All other building & civil engineering crafts not shown above total 61·8 × 1000

All other occupations not shown above total 54·5 × 1000

Fig. 4.6 (based on information compiled by the Ministry of Public Building and Works)

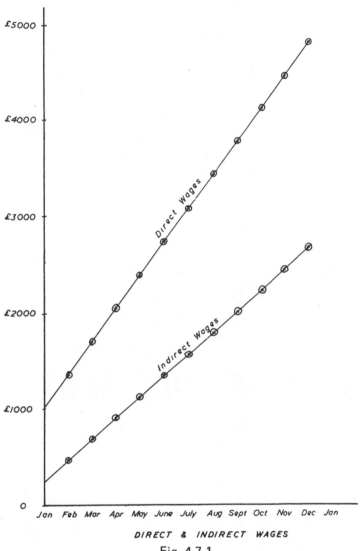

DIRECT & INDIRECT WAGES

Fig. 4.7.1

In amount-of-change charts the information is plotted on arithmetic scales, the charts being used to contrast different values. Fig. 4.7.1 records the cumulative direct and indirect wages for a small department within a firm, from which it appears that the lines representing direct and indirect wages are diverging with a consequent change in the relationship between the two. It is the ratio of direct to indirect wages that is important to management, and as the amount-of-change chart does not give this information advantage should be taken of the rate-of-change chart, which plots the information on a semi-log chart field.

The rate-of-change chart is used to show whether two or more sets of values are maintaining a constant ratio with each other. If the rate of change is the same in all sets of values, then the lines will remain a constant distance apart, but the lines will either converge or diverge if the rate of change alters. An example of this technique is to be found in the control of the direct to indirect wages ratio (Fig. 4.7.2): it can be seen that the lines representing direct and indirect wages are parallel, which indicates a constant relationship between the two. Another use of the rate-of-change chart is in the comparison of small values with very large values on one chart. In Fig. 4.8 a comparison is being made between a maximum of about 31 000 two-bedroomed flats and a minimum of about 350 four-bedroomed flats. Clearly if this presentation had been attempted on arithmetic paper of the same size or any other reasonable size the minimum value of 350 would not have been noticeable.

Other examples of graphs are given in the following section.

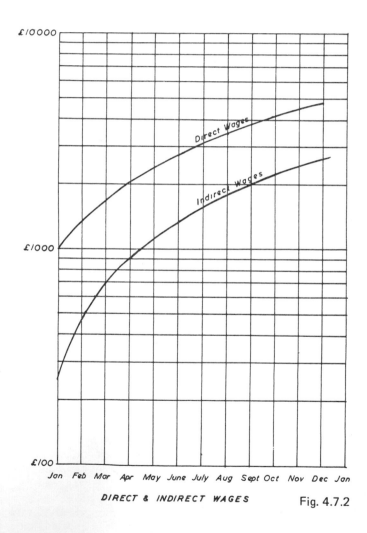

DIRECT & INDIRECT WAGES   Fig. 4.7.2

NEW FLATS COMPLETED FOR LOCAL AUTHORITIES 1965   Fig. 4.8 (based
Analysis by Bedroom Type   England & Wales   on Fig. 4.1)

4.3   USE OF STATISTICAL TECHNIQUES IN CONTROL AND FORECASTING

Statistical techniques are one of the tools of management, and one important use of these tools is in the control of a business undertaking. This control is achieved by comparing what was planned to happen with what is actually happening, thus ascertaining what variations and trends are occurring. Control is achieved by taking action to ensure that results which are not as planned are brought back on course as soon as possible; alternatively, the policy may need to be altered to suit the changing circumstances.

Forecasts can be made by observing present trends and tendencies and comparing these phenomena with past results, taking note of any cyclical trends that may occur. Whenever forecasts are made from statistical data, they should be tempered with the statistician's or manager's experience.

4.3.1 Time series methods

A time series is a number of values measured at particular points in time. Time series methods can be used either in forecasting the immediate future or alternatively for merely recording the past (a very useful exercise for the purpose of comparison). An example of a time series and its use in short term forecasting is shown in Fig. 4.9, which uses lunar months instead of calendar months so as to arrive at a more consistent time period. As the results of the sales of house type X become available month by month, they are plotted on the chart field using the monthly total scale, whilst at the same time the moving annual total can be plotted on the chart field using the moving annual total scale. The moving annual total is found by adding the present month's result to the previous twelve results (note that if this example had been based on calendar months the present month's result would be added to the previous eleven results). The advantage of the moving annual total is that it smooths out the variations that are occurring monthly, and indicates the trend of events. An example of this can be seen in Fig. 4.9, where although at the end of time period 5 there has been a considerable drop in sales it can nonetheless be seen from the moving annual total that the general trend of sales is still upwards.

From the trends indicated by the moving annual total it should be possible to forecast the result at the end of the year.

4.3.2 Z-chart

This chart is called the Z-chart because of the characteristic shape set up by the three graphs (monthly total, cumulative monthly total and moving annual total) plotted on the chart field.

This type of chart is particularly useful for recording sales statistics, but can be used to advantage with any data where trends are important. By plotting the budgeted sales in the form of a Z-chart and then plotting actual results on the chart as they become available, it is possible to depict discrepancies so that control can take place.

An example of a Z-chart is given for the sales of house type Y (Fig. 4.10). In this chart, as with the previous chart, it is possible to arrive at the trend in sales by plotting the moving annual total and thus smoothing out the monthly variations. It is common practice to plot the monthly figures to a larger scale than the cumulative and annual figures.

Fig. 4.9

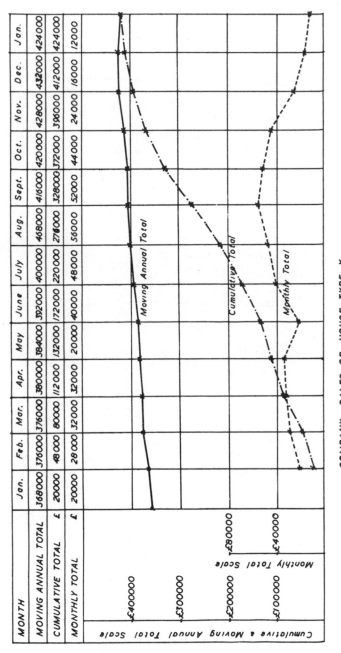

MONTH	Jan.	Feb.	Mar.	Apr.	May	June	July	Aug.	Sept.	Oct.	Nov.	Dec.	Jan.
MOVING ANNUAL TOTAL £	368000	376000	376000	380000	384000	392000	400000	468000	416000	420000	428000	432000	424000
CUMULATIVE TOTAL £	20000	48000	80000	112000	132000	172000	220000	276000	328000	372000	396000	412000	424000
MONTHLY TOTAL £	20000	28000	32000	32000	20000	40000	48000	56000	52000	44000	24000	16000	12000

COMPANY SALES OF HOUSE TYPE Y

Fig. 4.10

4.3.3 Relatives and index numbers

4.3.3.1 Relatives
The relative is a ratio between the value for any one year's result and the value for the standard year (usually called the base year), expressed as a percentage of the standard year. Relatives can be used for comparing changes in value of any variable, such as the volume of production or wage rates. As an example a list of relatives has been drawn up for the value of new orders obtained by contractors for new work in the housing field (Fig. 4.11); note that only the totals arrived at for each year are being considered.

Fig. 4.11 New housing orders obtained by contractors for new work in Great Britain (Based on the *Annual Digest of Statistics* [H.M.S.O., 1967])

Year	Public sector		Private sector	
	£ million	Relative	£ million	Relative
1961	273	100	469	100
1962	343	126	465	99
1963	454	166	526	112
1964	465	170	739	158
1965	510	187	693	148
1966	564	203	590	126

It can be seen from this chart that between the base year (1961) and 1962, housing in the public sector increased to 126 per cent, whereas housing in the private sector decreased to 99 per cent. In the year 1966 housing in the public sector had reached 203 per cent whereas housing in the private sector had only attained 126 per cent. It is difficult to compare the increases in work done in each sector without using relatives as both these series start at different levels.

4.3.3.2 Index numbers
Index numbers are a means by which the related movements of statistical variables can be estimated. They are used in cases where it would be difficult to measure actual movements and provide in a single term an indication of the variation against time of a group of related values. Index numbers are particularly useful because they indicate the trend.

If a contractor wanted to know how the cost of all the materials he used was varying from year to year, he could arrive at the answer by compiling index numbers for the cost of all these materials for each year and comparing them with some base year. The first step would be to ascertain the price of these materials in the base year and the relative importance of each of these materials (weight) in the contractor's work.

The index for the base year is given the value of 100, and a table is constructed (Fig. 4.12.1) to find the index number for the next year.

Fig. 4.12.1

Base year + 1

Material	Percentage rise in price over base year	Weight	Producer (weight × percentage rise)
A	5	25	125
B	17	14	238
C	13	10	130
D	20	12	240
E	11	8	88
F	9	5	45
G	10	20	200
		94	1066

Therefore the increase $= \dfrac{1066}{94} = 11$ per cent approximately.

Hence the index number for year x + 1 = 111.

This process would be repeated for each year as shown in Figs. 4.12.2, 4.12.3, and 4.12.4 so as to build up a list of index numbers.

Fig. 4.12.2

Year x + 2

Material	Percentage rise in price over base year	Weight	Producer (weight × percentage rise)
A	10	25	250
B	23	14	322
C	17	10	170
D	30	12	360
E	20	8	160
F	16	5	80
G	24	20	480
		94	1822

Therefore the increase $= \dfrac{1822}{94} = 19$ per cent approximately.

Hence the index number for year x + 2 = 119.

Fig. 4.12.3

Year x + 3

Material	Percentage rise in price over base year	Weight	Producer (weight X percentage rise)
A	11	25	275
B	24	14	336
C	18	10	180
D	25	12	300
E	25	8	200
F	15	5	75
G	20	20	400
		94	1766

Therefore the increase $= \dfrac{1766}{94} = $ 19 per cent approximately.

Hence the index number for year x + 3 = 118.

Fig. 4.12.4

Year x + 4

Material	Percentage rise in price over base year	Weight	Producer (weight X percentage rise)
A	12	20	240
B	24	14	336
C	18	10	180
D	26	12	312
E	27	8	216
F	17	5	85
G	21	20	420
		89	1789

Therefore the increase $= \dfrac{1789}{89} = $ 20 per cent approximately.

Hence the index number for year x + 4 = 120.

(Note that the weight of material A has changed, denoting a change in usage.)

The index numbers for the materials employed are therefore as follows:

$$\begin{aligned}
\text{Year } x &\quad 100 \\
\text{Year } x + 1 &= 111 \\
\text{Year } x + 2 &= 119 \\
\text{Year } x + 3 &= 119 \\
\text{Year } x + 4 &= 120
\end{aligned}$$

By building up these index numbers over a period of years it is possible to work out the trends in the cost of the materials used and thus to forecast the possible cost at some point in the future.

4.3.4 Quality control

4.3.4.1 Introduction

The aim of quality control is to ensure that all items produced can be used for their intended purpose by facilitating the elimination of defects and variations from the standards established for the production process. Quality control also seeks to avoid waste of time, materials and money by highlighting the point at which a production process is becoming defective, because if defective work did leave the factory and reach the site the cost of the consequent delays could be many times more than the cost of scrapping the original defective in the factory and providing a replacement. For example, if precast column units had to be altered on site, this would cause a hold-up in the production of the concrete frame which in turn would delay the frame. It may even be impossible to alter some units on site, which would necessitate remanufacture.

4.3.4.2 Statistical quality control

This is one of the most important control techniques based on the application of statistical theory. It may be applied to most processes that are of a repetitive nature, examples in the construction industry being the control of concrete and pre-cast concrete production, and the control of standard joinery in a joiners shop.

Statistical quality control is based upon the technique of statistical sampling.

When a production process is running normally, random samples of the items produced will be distributed within the limits of a normal frequency curve possessing the following properties

1. A mean value corresponding to the peak of the curve.

2. Three standard deviations either side of the mean containing 99.7 per cent of all the results.

3. Two standard deviations either side of the mean containing 95.4 per cent of all the results.

4. One standard deviation either side of the mean containing 68.2 per cent of all the results. (Fig. 4.13).

Any particular process can be represented by drawing up a frequency curve from random samples, and from this it is then possible to ascertain by plotting on any specific tolerance what proportion of the items produced have to be rejected. For example, if the tolerance coincided with three standard deviations to the lower side of the mean, then 0.15 per cent of all manufactured articles would be rejected. Consequently it should be possible, knowing the standards of performance obtainable in the manufacturing process and knowing the proportion of rejects that would be acceptable, to arrive at a combination of the two. This would then be the standard, i.e. the accepted process having a particular specification with an allowed number of rejects (one in x tested).

Once this information has been acquired, it is possible to set up a control chart (Fig. 4.14) to carry out progressive control of the process. Customarily the possible values are shown on the vertical axis and the sample number on the horizontal axis, whilst both the mean value and the control limits should also be indicated. A common setting for the upper and lower control limits is $\pm 3\sigma$,

Fig. 4.13

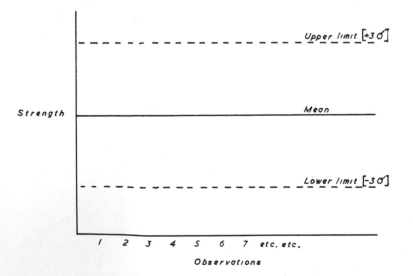

Fig. 4.14

but limits can be set at any level providing it is realized that the closer the limits are together, the greater will be the number of rejects.

As results become available they are plotted on the control chart, which throws into relief significant variations from the normal performance that require further investigation. Consistently high or low readings indicate faulty setting of the machine (e.g. if concrete was produced having a consistently high crushing strength, this may indicate that too much cement was being used in the batches, possible as a result of some fault in the cement weighing mechanism).

4.3.4.3 Quality control of concrete
Quality control can be carried out on either a retrospective or a progressive basis.

1. Retrospective quality control

When quality control is carried out on a retrospective basis at the end of a predetermined period or at the end of a project, the results are collected and analysed firstly in the form of a histogram and then by plotting these results on probability paper. The mean and standard deviation for this set of results can then be ascertained, and from the line on the probability paper it is also possible to find the probability of any result occurring. By knowing the standard of control for this set of results, it is therefore possible to undertake another project requiring the same standard of control with prior knowledge of the probability that a particular tolerance can be met.

Fig. 4.15.1

Cube Crushing Strength at 28 days in MN/m²							
1	20.52	26	18.92	51	26.32	76	22.45
2	29.15	27	14.98	52	28.37	77	26.53
3	24.16	28	20.33	53	20.17	78	24.59
4	28.25	29	18.66	54	30.92	79	25.13
5	18.32	30	17.84	55	27.63	80	27.51
6	29.22	31	20.78	56	25.78	81	23.77
7	26.35	32	19.43	57	26.77	82	32.51
8	29.11	33	15.66	58	31.67	83	24.22
9	22.22	34	21.77	59	22.88	84	28.39
10	30.12	35	18.82	60	31.53	85	22.32
11	27.47	36	16.97	61	24.12	86	33.42
12	19.54	37	21.96	62	29.88	87	25.26
13	30.87	38	19.75	63	27.41	88	34.66
14	33.68	39	22.72	64	31.44	89	24.37
15	23.66	40	20.94	65	22.73	90	29.75
16	30.66	41	24.81	66	30.81	91	35.51
17	33.75	42	17.23	67	20.98	92	22.56
18	19.80	43	26.22	68	28.49	93	28.95
19	32.16	44	20.87	69	30.67	94	34.72
20	27.90	45	27.81	70	22.61	95	25.77
21	35.22	46	17.49	71	29.48	96	35.42
22	25.38	47	26.96	72	32.62	97	23.52
23	37.37	48	23.67	73	26.67	98	36.78
24	25.26	49	24.75	74	33.87	99	37.67
25	26.86	50	21.58	75	27.92	100	24.11

CLASS	FREQUENCY			
	1st 25 results		Total 100 results	
14.00-15.98			II	2
16.00-17.98			IIII	4
18.00-19.98	III	3	⫴⫴ III	8
20.00-21.98	I	I	⫴⫴ ⫴⫴	10
22.00-23.98	II	2	⫴⫴ ⫴⫴ II	12
24.00-25.98	III	3	⫴⫴ ⫴⫴ IIII	14
26.00-27.98	IIII	4	⫴⫴ ⫴⫴ ⫴⫴	15
28.00-29.98	IIII	4	⫴⫴ ⫴⫴ I	11
30.00-31.98	III	3	⫴⫴ IIII	9
32.00-33.98	III	3	⫴⫴ II	7
34.00-35.98	I	I	⫴⫴	5
36.00-37.98	I	I	III	3
	TOTAL 25		TOTAL 100	

Fig. 4.15.2

As an example, assume that the results shown in Fig. 4.15.1 are the results obtained on a particular project. The steps to be taken in analysing these results are as follows:

A. Compile a frequency distribution chart (Fig. 4.15.2), which will indicate the number of results occurring in a particular class.

B. Plot these results in the form of a histogram (Fig. 4.15.3), which gives a clearer impression of the distribution of results (the histogram is based on results recorded in the frequency distribution chart). It is helpful for the next stage if a table of the cumulative frequency and cumulative frequency percentage is recorded with the histogram.

C. Construct a probability graph. This is a straight line graph drawn on probability paper (Fig. 4.15.4), the horizontal scale being divided into the class intervals and the vertical scale into percentages. The cumulative percentages are plotted on the chart field in the appropriate class intervals, and a straight line is drawn through these points. If a perpendicular is dropped from where this line crosses the 50 per cent mark down to the horizontal axis, it will cut the horizontal axis at the mean value, and if lines are dropped from where the inclined line cuts the 16 per cent and 84 per cent lines, these lines will cut the horizontal axis at one standard deviation at either side of the mean (1σ either side of the mean contains approximately 68 per cent of all the results). Consequently we have now obtained the mean and the standard deviation for this series, and from the chart it is also possible to ascertain the probability of achieving a particular result: if the desired minimum crushing strength is 18.00 MN/m^2, for example, then by drawing a perpendicular line up from the horizontal axis at this point until it cuts the inclined line it can be seen that there is a 9 per cent chance of falling below this limit or alternatively a 91 per cent chance of exceeding the limit. If this rate of failure is acceptable in the new process, then the standard of control exercised before will suffice; if not, then tighter control will be necessary.

As an alternative to finding the standard deviation and the mean by this method, it is possible to calculate them. First the information shown in the frequency distribution table (Fig. 4.15.2) is analysed into the following form:

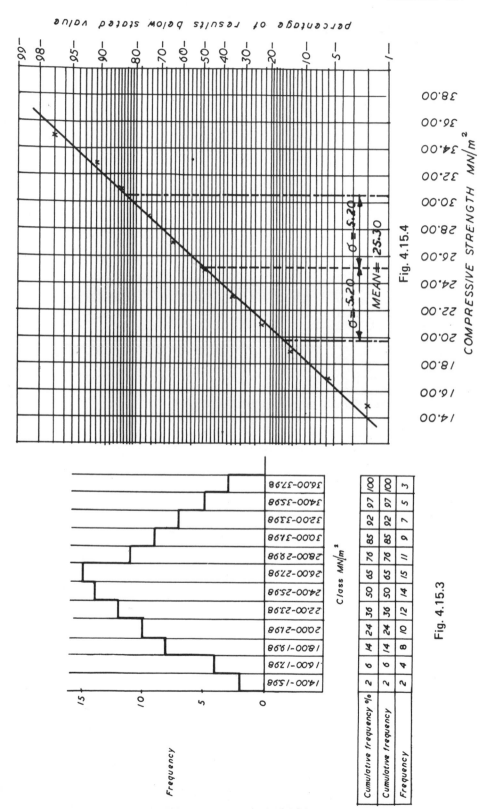

Fig. 4.15.4

Fig. 4.15.3

Class MN/m²	14.00-15.98	16.00-17.98	18.00-1.998	20.00-21.98	22.00-23.98	24.00-25.98	26.00-27.98	28.00-29.98	30.00-31.98	32.00-33.98	34.00-35.98	36.00-37.98
Cumulative frequency %	2	6	14	24	36	50	65	76	85	92	97	100
Cumulative frequency	2	6	14	24	36	50	65	76	85	92	97	100
Frequency	2	4	8	10	12	14	15	11	9	7	5	3

Fig. 4.15.5

x	f	u	uf	u^2f	fx
15.00	2	−6	−12	72	30
17.00	4	−5	−20	100	68
19.00	8	−4	−32	128	152
21.00	10	−3	−30	90	210
23.00	12	−2	−24	48	276
25.00	14	−1	−14	14	350
assumed mean 27.00	15	0	0	0	405
29.00	11	1	11	11	319
31.00	9	2	18	36	279
33.00	7	3	21	63	231
35.00	5	4	20	80	175
37.00	3	5	15	75	111
	100		−47	717	2606

Where: x is the class mid-point
f is the frequency
u is the number of classes this class is removed from the assumed mean
uf is the column u multiplied by the column f
u^2f is the column u squared and multiplied by the column f
fx is the column f multiplied by the column x

$$\text{True mean} = \text{assumed mean} + c \times \frac{\Sigma uf}{\Sigma f}$$

where c = class interval

$$\therefore \text{True mean} = 27.00 + 2 \times \frac{-47}{100}$$

$$= 27.00 - 0.94$$

$$= 26.06 \text{ MN/m}^2$$

$$\text{standard deviation } \sigma = c \times \sqrt{\frac{\Sigma u^2 f}{\Sigma f} - \left(\frac{\Sigma uf}{\Sigma f}\right)^2}$$

$$= 2.00 \times \sqrt{\frac{717}{100} - \left(\frac{-47}{100}\right)^2}$$

$$= 2.00 \times \sqrt{7.17 - .22}$$

$$= 2.00 \times \sqrt{6.95}$$

$$= 5.25 \text{ MN/m}^2$$

It will be noticed that these values do not agree with the results arrived at by graphical methods, but this is only to be expected.

Fig. 4.16.1

Proportion falling below lower control level	K
1 in 10	1.28
1 in 20	1.64
1 in 25	1.75
1 in 33	1.88
1 in 40	1.96
1 in 50	2.05
1 in 100	2.33

2. Progressive quality control

If quality is controlled on a progressive basis, then it is possible to distinguish adverse trends at an early stage and control them by taking corrective action.

To perform this type of control, a chart is drawn up (Fig. 4.14) with concrete strength on the vertical axis and the result number on the horizontal axis. The mean strength and the control levels must now be plotted on the chart, these values being obtained in two different ways.

A. It is quite possible that the designer of the concrete mix made a number of assumptions about the value of the standard deviation and the mean strength, knowing the standard of control that would be possible on site. He will have specified the minimum strength on this basis, and in addition he will know what proportion of the results can be allowed to fall below a certain level. Consequently the anticipated mean strength may be plotted on the chart. The control levels can also be plotted on at $\pm K\sigma$ from the mean where K is a constant from a table (Fig. 4.16.1) and σ is the anticipated standard deviation.

B. The other possibility is that the standard of control is not known. In this case the standard deviation is ascertained after approximately 25 cubes have been taken at random and tested. To find the mean strength necessary to ensure that no more than a specified number of results fall below the specified minimum, the standard deviation σ must be found and $K\sigma$ added to the minimum strength where K would have some desired value (i.e. 1 in 100, K = 2.33; 1 in 25, K = 1.75). This value is then plotted on the chart, after which the control levels are also entered by using the same method. It is common practice to repeat this process when more results become available. As an example in which the mean and standard deviations are not known, assume that the data given in Fig. 4.15.1 relates to a project that is being controlled progressively, so that the first 25 results can be considered as the results for the trial period. The mean and standard deviation must first be found for these results by following the method used for retrospective control (Figs 4.16.2 and 4.16.3). From Fig. 4.16.3 it can be seen that the mean is 26.30 MN/m² and the standard deviation is 5.20 MN/m². These results are now used to construct the control chart (Fig. 4.16.4) as follows:

(a) Draw out the axes, with the horizontal representing the test cube number and the vertical the compressive strength

(b) Insert the minimum strength ón the chart (assume this to be 16.50 MN/m²).

(c) Assuming an allowable failure rate of 1 in 100, the required mean strength can be calculated from mean strength = $K\sigma$ + minimum strength, where σ is the standard deviation of the first 25 results (Fig. 4.16.3).

Hence mean strength = 2.33 x 5.20 + 16.50
= 12.11 + 16.50
28.61 MN/m²

(d) Plot the calculated mean strength on the control chart. If the calculated mean strength is compared with the mean strength for the first 25 results it can be seen that the achieved mean strength is less than that required, so it can be assumed that the control is not good enough. Therefore either the control would have to be improved or the mix design would have to be altered to ensure that only one result in 100 exceeded the limits set.

(e) As more test results become available they are plotted on the chart (Fig. 4.16.5) and if the number of results occurring on either side of the mean is counted it soon becomes evident should the average strength of the concrete being produced fall below the required average. It is also possible to get an impression of the variability after approximately 10 results have been plotted by checking the number of results falling outside the control limits for 1 in 10. When another 10 results are available they can also be checked against the 1 in 10 control limits, and if they are added to the previous 10 results they can then be checked against the 1 in 20 control limits. If more results than planned fall outside these limits then it is necessary to tighten up control to reduce the standard deviation.

This example is based on the concept of using individual cube samples as test data. A more sophisticated concept is the use of groups of cubes as samples

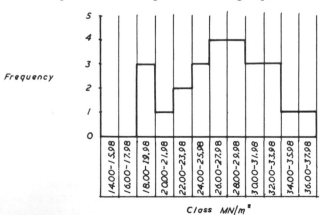

Cumulative frequency %			12	16	24	36	52	68	80	92	96	100
Cumulative frequency			3	4	6	9	13	17	20	23	24	25
Frequency			3	1	2	3	4	4	3	3	1	1

Fig. 4.16.2

rather than individual cubes. In the latter case it is necessary to use two control charts—one to control the range and the other to control the mean of the cube sample.

Fig. 4.16.3

Fig. 4.16.4

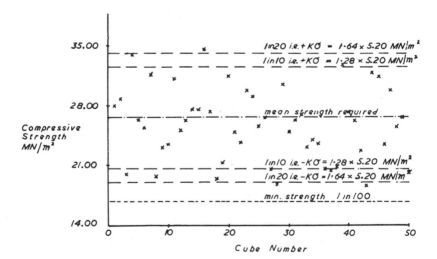

Fig. 4.16.5

5: Budgetary and Cost Control

5.1 BUDGETS AND BUDGETARY CONTROL

5.1.1 Introduction
Budgets have been used for planning the income and expenditure of a country's government for many years. With the growing complexity of the building industry it is essential that firms also plan their policy well into the future, and it is through budgets that such plans can be converted into the quantitive and monetary terms which are a company's fundamental objectives.

5.1.2 Policy budgets in the construction industry
By compiling budgets a firm can arrive at a selling or tendering price for its projects, having considered what resources are available to carry out the work. The firm will need to know, for example, how much capital is necessary, and having acquired this information it can then decide whether to expand or contract any particular aspect of the business, bearing in mind competition and demand. From past results it may also be possible to determine a change in demand that could give an indication of possible new expansion plans.

5.1.3 Project budgets
It is important for any contractor to know what capital is going to be needed for a project and when it will be required. In order to ascertain this, the contractor will have to draw up a programme for the project and from this find the rate of expenditure and rate of income, based on a particular time period. The difference between these two will give the amount of capital required in this time period (examples are given later in this chapter).

If required, budgets can be prepared for the labour to be used (based on the labour graph), for materials and plant (based on schedules and the programme), and for the site overheads (based on the programme).

5.1.4 Achievement of budgets
Once a budget has been established, the next objective is to achieve the target set. This is attained through a system of budgetary control. Like any other control technique, budgetary control is a continuous comparison of the actual achievement with that planned, which highlights any deviations from the plan so that action can be taken either to bring things back on course or to change the plan.

5.1.5 Examples of budgets and budgetary control
Two examples will be given in the following pages, one of which is a non-repetitive construction based on the amenity centre (chapter 1, section 1.2) whilst the other is an example of repetitive construction based on the shells (chapter 1, section 1.7.3). It must be emphasized that these examples are very much simplified and are put forward for illustrative purposes only.

5.1.5.1 Non-repetitive construction
It is assumed that a contract has been entered into with the client and that

payment for work carried out will be made 14 days after the end of the month in which the work is completed. Retention will be 10 per cent initially with a maximum retention of 5 per cent of the contract sum (£80790). Half the retention will be released 14 days following the completion of the project, the other half will be released six months after completion.

The cost per operation has been ascertained and allocated to the master construction programme. From this programme it is now possible to arrive at the expenditure for each month (Fig. 5.1.1).

Fig. 5.1.1

	Own expenditure (£)	Payments to sub-contractors (£)	Cumulative expenditure (£)
Week 4	4220		4220
Week 8	6340		10560
Week 12	6440		17000
Week 16	7380		24380
Week 20	5160		29540
Week 22		540	30080
Week 24	3230		33310
Week 26		2700	36010
Week 28	6980		42990
Week 30		2574	45564
Week 32	9870		55434
Week 34		6462	61896
Week 36	3140		65036
Week 38		3222	68258
Week 40	2730		70988
Week 42		1582	72570

Using a similar method, it is possible to arrive at the programme of income for the project, the value of each operation having been ascertained from the bill of quantities. The expected amounts of the monthly valuations are shown in Fig. 5.1.2.

Fig. 5.1.2

	Own income (£)	Retention (£)	Income less retention (£)	Payments for sub-contractor (£)	Cumulative income (£)
Week 6	4590	459	4131		4131
Week 10	6890	689	6201		10332
Week 14	7000	700	6300		16332

Week 18	8020	802	7218		23850
Week 22	5660	566	5094	600	29544
Week 26	3780	378	3402	3000	35946
Week 30	7840	406	7434	2860	46240
Week 34	11320	4000 limit of retention	11320	7180	64740
Week 38	3620		3620	3580	71940
Week 42	3090		3090	1760	76790
		2000 retention repaid			78790
Week 46					
Week 50					
Week 54					
Week 58					
Week 62					
Week 66		2000 retention repaid			80790

(N.B. Payment takes place 14 days after the end of the month in which the work is carried out.)

These figures for income and expenditure have been recorded graphically in Fig. 5.1.3. The capital required at any time can now be ascertained by finding the difference between income and expenditure.

5.1.5.2 Repetitive construction

It is assumed that a contract has been entered into with the client and that payment will be made in stages for each unit, e.g. excavation, concrete foundations, brickwork, roof carcass, and roof finishes. Payment will be received one week after completion of a stage, and the contract makes no provision for retention money.

It has been ascertained that the total cost of completing these shells is £13655, which is built up in the following manner:

Excavation	£165	=	£16.5 each block
Concrete foundations	£3190	=	£319 each block
Brickwork	£6340	=	£634 each block
Roof construction	£1320	=	£132 each block
Roof finish	£2640	=	£264 each block
	£13655		

By allocating these costs to the programme of work it can be seen that the distribution of cost will be that shown in Fig. 5.2.1.

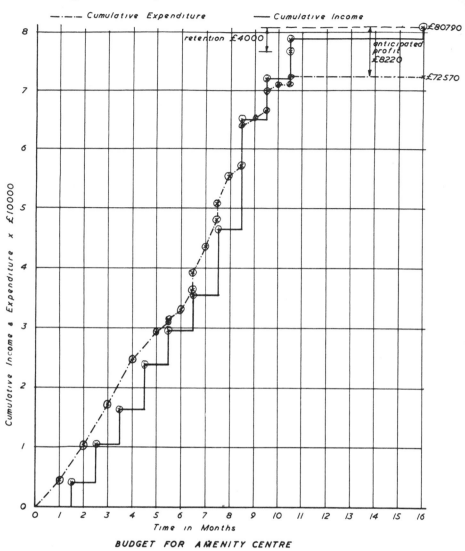

BUDGET FOR AMENITY CENTRE

Fig. 5.1.3

Fig. 5.2.1	Weekly expenditure £	Cumulative expenditure £
Week 1	280.5	280.5
Week 2	1074.0	1354.5
Week 3	1463.5	2818.0
Week 4	1727.5	4545.5
Week 5	1686.25	6231.75
Week 6	1686.25	7918.0
Week 7	1686.25	9604.25
Week 8	1686.25	11290.5
Week 9	1447.0	12737.5
Week 10	653.5	13391.0
Week 11	5264.0	13655.0

(N.B. In practice these figures would be taken to the nearest £10 but here precise calculations have been made so that the reader can follow the working.)

The contract sum is £15710 and the value of each stage (as ascertained from the priced bill of quantities) is as follows:

Excavation	£190	=	£19 each block
Concrete foundations	£3670	=	£367 each block
Brickwork	£7290	=	£729 each block
Roof construction	£1520	=	£152 each block
Roof finish	£3040	=	£304 each block
	£15710		

By allocating these values to the programme of work it can be seen that the distribution of the value will be as shown in Fig. 5.2.2 (only completed stages will be measured at the end of the week).

Fig. 5.2.2

	Weekly income (£)	Cumulative income (£)
Week 1	0	0
Week 2	38	38
Week 3	1520	1558
Week 4	1286	2844
Week 5	1609	4453
Week 6	1552	6005
Week 7	3104	9109
Week 8	1552	10661
Week 9	1552	12213
Week 10	1552	13765
Week 11	1641	15406
Week 12	4304	15710
	15710	

Both sets of figures can be plotted on a chart field (Fig. 5.2.3) and by subtracting the income from the expenditure the capital required at that particular time can be found.

In both of the foregoing examples the plotting of the actual income and expenditure on the charts as soon as figures become available highlights descrepancies and enables control to be exercised.

5.2 COST CONTROL

5.2.1 Introduction

Costing has been carried out in the building industry for a number of years, but the type of costing used by some firms only gives the total cost of the project when it has been completed. This is then compared with the amount received, but if the project has lost money there is little that can be done about it. It is obviously a distinct advantage if a contractor can ascertain which section of a project is in deficit and know when it starts losing money. Cost control techniques have been developed for this purpose.

5.2.2 Objectives

The main objectives of cost control are:

1. To see that the company's policy with regard to production is carried out, which in turn will ensure that planned profit margins are maintained.

2. To arrive at the cost of each stage, operation (in the case of repetitive construction), or unit, and to carry out a continuous comparison with the target to ascertain the gain or loss on each. This information must be available early enough for corrective action to be taken.

3. To provide information on cost for use in future estimating.

BUDGET FOR HOUSE SHELLS

Fig. 5.2.3

5.2.3 Cost control systems

Before deciding upon the degree of sophistication of a system, consideration must be given to the cost of operating it and the benefits it provides. Clearly a balance should be reached between the two.

Systems can vary from those which control the work on a section or stage basis to those which control it on a unit basis, and in some systems only certain sections of the project are selected for control. It is usual to limit site cost control to the control of labour and plant as these are the areas where there is likely to be the greatest amount of variability.

5.2.4 Timing of presentation of cost information

When cost control techniques are used the system must be administered in such a manner that the site manager is made aware of any deviations from planned cost as soon as they are recognized. Ideally the costs should be checked daily, but on most building operations this would be extremely expensive and the cost of the work done is in consequence usually checked weekly.

5.2.5 Uses of cost information

Cost control information is invaluable for future estimating. Feed back must be accompanied by a full description of the conditions under which the work was carried out, as conditions can vary a great deal from one contract to another.

Another very important use of cost control information is in the pricing of variations. If the variations can be identified before the work commences and a separate record is kept of work costs on them, the contractor will have factual information to assist him in settling a rate for the work done.

The weekly measure carried out for the cost control system can also be used for the monthly valuation and thus eliminates the need for the work to be re-measured at the end of each month, although materials on site would have to be a ascertained separately for the purpose of the monthly valuation.

5.2.7 Unit cost control

In this chapter it is proposed to deal with only two systems of cost control. The first system is one which could be applied to non-repetitive construction and is usually referred to as a unit cost control system. In the example given, this has been applied to the amenity centre (chapter 1, section 1.2).

It has already been seen that for unit costs to be of value as a control tool they must be issued as soon as possible after their compilation, which is usually the week following that to which they refer. Any unit costs in excess of or below the target by more than a predetermined margin require further investigation and should be brought to the attention of management, thus applying the principle of management by exception. The information is collected by means of the allocation sheet and the weekly summary sheet, together with actual measurement of work done.

5.2.7.1 Allocation sheet (Fig. 5.3)

This is the daily record of the hours worked by an operative or gang from which the hours spent on an operation can be obtained.

The sheets are the basis of the cost control system and must therefore be accurately compiled by the man in charge of the gang. To achieve this accuracy,

DAILY ALLOCATION SHEET												
CONTRACT						FOREMAN/GANGER						
CONTRACT NO.						TRADE						
DATE						SHEET NO.		OF				

*BLOCK NO. & OPERATION / NAME & CLOCK NO.

	T	L	T	L	T	L	T	L	T	L	DAILY TOTAL FOR OPERATIVE	STOPPAGES [HOURS]
TOTALS												

PLANT	PLANT NO.	O T	S T	O T	S T	O T	S T	O T	S T	O T	S T	DAILY TOTALS FOR PLANT	
												OPERATING	STANDING
TOTALS													

REMARKS

WEATHER CONDITIONS   SIGNED

FOREMAN/GANGER

* BLOCK NO. ONLY APPLIES WHEN THIS FORM IS BEING USED FOR HOUSING

Fig. 5.3

WEEKLY   SUMMARY   SHEET

CONTRACT
CONTRACT NO.
＊BLOCK NO.

WEEK NO.   WEEK ENDING
SHEET NO.   OF
COMPILED BY

OPERATION :-

PLANT USED :-

DAY	M	TU	W	TH	F	SAT	SUN	TOTAL
LAB. HRS.								
TRADE. HRS.								
PLANT HRS.								

OPERATION :-

PLANT USED :-

DAY	M	TU	W	TH	F	SAT	SUN	TOTAL
LAB. HRS.								
TRADE. HRS.								
PLANT. HRS.								

OPERATION :-

PLANT USED :-

DAY	M	TU	W	TH	F	SAT	SUN	TOTAL
LAB. HRS.								
TRADE. HRS.								
PLANT. HRS.								

OPERATION :-

PLANT USED :-

DAY	M	TU	W	TH	F	SAT	SUN	TOTAL
LAB. HRS.								
TRADE. HRS.								
PLANT. HRS.								

OPERATION :--

PLANT USED :-

DAY	M	TU	W	TH	F	SAT	SUN	TOTAL
LAB. HRS.								
TRADE. HRS.								
PLANT. HRS.								

＊BLOCK NO. ONLY APPLIES WHEN THIS FORM IS BEING USED FOR HOUSING

Fig. 5.4

they should be completed daily and handed in to the site office where the operation descriptions, time per operation and total time per operation will be checked.

When plant is involved in operations, this should be recorded by the plant operator if the machine has a permanent operator and does not work with a gang, e.g. the excavator. If the item does not have a permanent operator (for example the vibrating roller) or always works in conjunction with a gang (the concrete mixer), then it should be booked by the ganger.

Remarks should be made about working conditions, any special methods used, etc. Note should also be made of how non-productive time is made up (i.e. how much is standing time and how much breakdown time).

5.2.7.2 Weekly summary sheet (Fig. 5.4)

This sheet is used to collect all the information on operational hours from the daily allocation sheets, and will be compiled by the cost surveyor at the end of each week. The hours recorded on the allocation sheets are ordinary hours and net overtime spent on operations, so that the weekly summary sheet presents the total basic hours spent on an operation in that week. This information is required for calculating the amount of bonus payable as well as for insertion in the weekly cost sheet.

5.2.7.3 Measurement of work done

Measurement of work done should be carried out by a person who fully understands the demarcation between operations so that the measure can be directly related to the costs arrived at through the allocation sheets and summary sheets.

In the type of work involved in the amenity centre many of the operations will be non-repetitive and these will have to be measured physically by tape and/or from working drawings. A record of the weekly measure should be kept in the form of record drawings which will be marked up weekly with the amount of work done. This record will be particularly useful when bonus is paid, ensures that gangs are not paid twice for the same piece of work, and constitutes a detailed record of progress. Items such as columns can be measured by simply observing their location and the number completed, the quantities being worked out from the drawings.

5.2.7.4 The weekly cost control sheet (Fig. 5.5)

This sheet is used for the collection of all the information arrived at via the allocation sheets, summary sheets, and records of measure. The unit cost for each operation can then be calculated and this facilitates control by comparison with set unit targets.

Column A and B

These columns list the operation numbers and operation description for the work carried out during the week in question.

Column C

This column records the amount of work carried out on each operation.

Column D

This column records the value of one unit of measure and would be set by the production control department. Indirectly, there would be a tie-up between these target values and the bill of quantities: for a section of work.e.g. brickwork, the total values based on the target set can be compared with the total of the relevant items in the bill of quantities and on completion of each section the

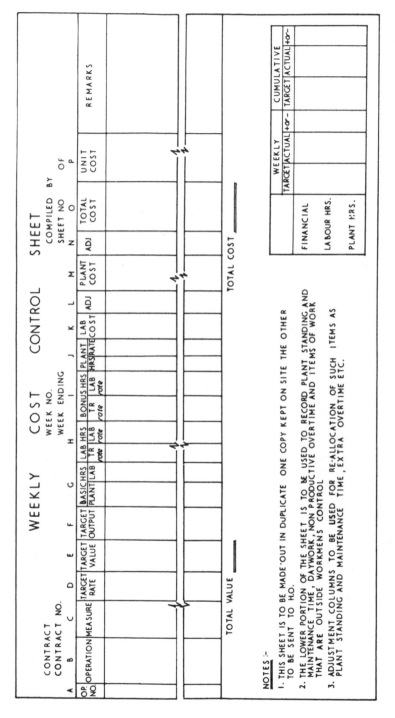

Fig. 5.5

total cost is compared with the bill of quantities to give a guide for future estimating.

Column E

This column records the target value of work carried out on one individual operation and the total of this column gives the total value of work done for the week, which can later be compared with the cost. The individual values are arrived at by multiplying the measure in column C by the rate of column D.

Columns F and G

Column F lists the target output for each unit of measure. By multiplying this by the measure (column C) the basic labour- or plant-hours can be calculated for insertion in column G.

Column H

This column records the actual labour-hours arrived at via the weekly summary sheet. The hourly rate paid to the workmen is recorded to facilitate the calculation of the operation labour cost.

Column I

The hours inserted in this column are the target hours (column G) less the actual hours (column H). The rate paid will be a proportion of the workmen's hourly rate, dependent on company policy.

Columns J and M

Column J records the plant-hours for the operation. If this is multiplied by the rate, the plant cost can be calculated and recorded in column M.

Column K

This column records the total labour cost before adjustment is calculated by adding the costs arrived at via columns H and I.

Column L

This column is used for the re-allocation on a *pro rata* basis of items such as non-productive overtime that have been recorded on the lower section of the sheet.

Column N

This column is used for the re-allocation on a *pro rata* to hours basis of items such as plant standing and plant maintenance. When plant breaks down there should be no hire charge to the site, as the cost of the breakdown is borne by either the plant department or the plant hire firm.

Column O

This is the total weekly cost for an operation and is arrived at by adding columns K, L, M and N. By adding up the values given in this column, the total cost for the week is arrived at.

Column P

This records the actual cost of producing one unit of measure on an operation, e.g. the cost of one cubic yard of concrete. It is arrived at by dividing the total cost for an operation (column O) by the total measure for that operation (column C); this unit rate can then be compared with the target rate in column D.

5.2.8 Operational cost control

The second system of cost control can be applied to repetitive construction and is usually classed as an operational cost control system. In the example it has been applied to the housing site at Tenbury (chapter 1, section 1.7.4).

The object of operational cost control is to arrive at the cost of completing each operation on each block and then to compare this with a target for each. Any operational cost in excess of or below the target by more than a predetermined margin requires further investigation and should be brought to the attention of management. As with unit cost control, information is

A	B		C	D		E		F	G	H	I	J	K

WEEKLY   OPERATIONAL   COST   SHEET

CONTRACT

CONTRACT NO.

WEEK NO.   WEEK ENDING

SHEET NO.   OF.

COMPILED   BY

BLOCK NO.	OPERATION		BONUS TARGET HRS.	ACTUAL LAB.HRS.		BONUS HRS.		PLANT	LAB. COST	ADJ. COST	PLANT COST	ADJ. COST	TOTAL COST	REMARKS
	OP. NO	OP.		TR. rate	LAB. rate	TR. rate	LAB. rate	HRS	RATE					

NOTES

1 THE LOWER PORTION OF THE SHEET IS TO BE USED TO RECORD PLANT STANDING AND MAINTENANCE TIME, DAYWORK, NON PRODUCTIVE OVERTIME AND ITEMS OF WORK THAT ARE OUTSIDE WORKMENS CONTROL

2 ADJUSTMENT COLUMNS TO BE USED FOR RE-ALLOCATION OF SUCH ITEMS AS PLANT STANDING AND MAINTENANCE TIME, EXTRA OVERTIME ETC,

3 ONLY COMPLETED OPERATIONS ARE TO BE ENTERED ON THIS SHEET

Fig. 5.6

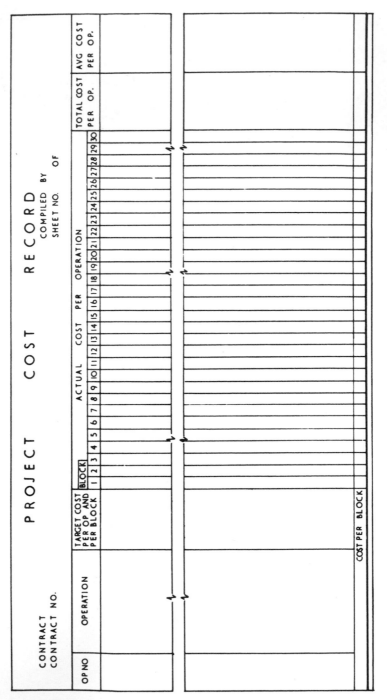

Fig. 5.7

collected by means of daily allocation sheets, weekly summary sheets, and actual measurement of the work done.

5.2.8.1 Daily allocation and weekly summary sheets
The daily allocation sheets and weekly summary sheets for this system will be similar to those for the unit cost control system (Figs 5.3 and 5.4).

5.2.8.2 Measurement of work done
On repetitive construction it is not necessary to measure the work done by taking physical dimensions on site. Instead, the project is divided into repetitive operations of known work content (ascertained from drawings) and the measurement of work done is carried out by observing the number of operations completed and the block number where these observations were made. The exception to this rule is with work below the damp-proof course, where variations in the work content may occur from one block to another, necessitating physical measurement of the work.

5.2.8.3 Weekly operational cost sheet
This sheet is used to arrive at the cost of completed operations, and is a means of converting the information presented in the weekly summary sheet into monetary terms. It is very similar to the unit cost sheet with the exception of columns A and C: column A simply records the block number, whilst column C lists the target hours for bonus purposes.

Only completed operations are entered on the operational cost sheet, the hours spent on uncompleted operations being retained on the weekly summary sheet and carried forward to the next weekly summary sheet. When these operations are completed the hours are transferred first to the operational cost sheet and then to the project cost record.

5.2.8.4 Project cost record
This sheet records the cost of every operation completed to date. As information becomes available it is added to the operational cost sheet week by week, so that this sheet enables the cost of completing each operation and each block to be readily compared with the original estimates.

6: Operational Research

6.1   INTRODUCTION

Like the other management techniques mentioned in this book, Operational Research is another example of the application of scientific method in the solution of management problems.

Decisions ultimately made by management are taken after consideration of relative advantages and disadvantages of various courses of action.

Small problems can be solved manually as shown in the examples which follow, but for more complex problems computers are far more economical and are of course faster and more accurate. Many standard computer packages are available in which case, once the problem has been formulated, the provision of the solution is a relatively simple process.

When using Operational Research; a scientific model, i.e. a set of equations, or a computer model, etc., is developed, and by applying data, decisions can be made more realistically. Operational Research is therefore dynamic.

6.1.1 Steps taken in applying operational research

The steps taken in applying Operational Research are:—

1. Define the problem.
2. Collect and record all relevant data.
3. Examine and analyse data and develop a mathematical or systematic model of the real life situation.
4. Check that model is valid.
5. Test model under differing conditions.
6. Select the optimum solution using the model.
7. Implement the results.
8. Keep a check on the model and see it is still valid under changing circumstances.

It can be seen that these steps are very similar to those used in Work Study.

It very often happens that Operational Research will not give a single optimum solution and the manager's judgement will be necessary to decide what action should be taken.

6.2   LINEAR PROGRAMMING

6.2.1 Introduction

Linear programming is probably the most used of the mathematical methods. It can be used for solving many problems, examples being:—

1. Optimising use of resources.
2. Determining most satisfactory product mix.
3. Division of work between different units.
4. Transportation problems.
5. Personnel assignment.
6. Establishing best processing schedules.
7. Determining optimum size of bid.
8. Location of new production plants, offices and warehouses.

The problem is to minimise or maximise some particular feature, this may be maximising profit or minimising loss, etc.

Problems containing only two variables can be solved graphically. Problems with three variables can also be solved graphically, if a three dimension graph can be drawn, but this can be very complicated. With more than three variables, a technique known as 'Simplex' is used and this is the most used method of solution.

6.2.2 Example of use of linear programming

A builder has a hydraulic press for producing paving slabs. The press is capable of producing 125 slabs per day. He has just bought some land on which to manufacture unpressed precast slabs but due to limitation in space, he can only produce 200 precast slabs per day.

The builder employs 18 men on this type of operation and does not wish to increase his labour force. 8 man days are required to produce 100 precast slabs and 4 man days for 100 hydraulically pressed slabs.

It is calculated that the profit will be £8 per 100 for precast concrete and £6 per 100 for hydraulically pressed slabs.

The problem is to maximise total profit.

6.2.3 The solution

Let S_1 be daily production of precast slabs (in hundreds).

Let S_2 be daily production of hydraulically pressed slabs (in hundreds).

This is a Linear Programming problem with two decision variables, S_1 and S_2. The builder can produce as many of each as he likes. If he makes S_1 precast slabs and S_2 hydraulically pressed slabs he will make a daily profit of

$$8S_1 + 6S_2.$$

This is known as the objective function and will give the total profit which he wishes to maximise.

The maximum number of each type which can be produced is 200 and 125 respectively.

∴
$$S_1 \leqslant 2$$
$$S_2 \leqslant 1\tfrac{1}{4}$$

There is also a side constraint because of the limit in the labour available.

The maximum labour available is 18 men, and it takes 8 man days to produce 100 precast slabs and 4 man days to produce 100 hydraulically pressed slabs. Therefore the total programme requirements are

$$8S_1 + 4S_2 \text{ man days}$$

and
$$8S_1 + 4S_2 \leqslant 18$$

NOTE: if he wished to produce 200 precast and 125 hydraulically pressed slabs, this would require

$$8 \times 2 + 4 \times 1\tfrac{1}{4} = 21 \text{ man days}$$

and this is not feasible due to labour force constraint. It is of course possible to produce 175 precast and 100 hydraulically pressed slabs. This would keep all labour occupied and would leave excess capacity on presses and in the yard.

$$(8 \times 1\tfrac{3}{4} + 4 \times 1 = 18)$$

The profit in this case would be

$$8 \times 1\tfrac{3}{4} + 6 \times 1 = £20$$

6.2.3.1 *Graphical solution*

As there are only two decision variables, the problem can be represented graphically. The area of possible production programme is limited by the capacity of the presses (125 slabs per day) and the space available for precast slabs (200 slabs per day). See Fig. 6.1.

Fig. 6.1

Commonsense dictates that $S_1 \geqslant 0$ and $S_2 \geqslant 0$

Fig. 6.2 shows the effect of the constraint

$$8S_1 + 4S_2 \leqslant 18$$

As the relationship is linear, two points are sufficient to plot the graph. The easiest procedure is to determine the value of S_1 when $S_2 = 0$ and the value of S_2 when $S_1 = 0$.

Let
$$S_2 = 0$$
$$8S_1 + 0 = 18$$
$$S_1 = \tfrac{18}{8} = 2\tfrac{1}{4}$$

Let
$$S_1 = 0$$
$$0 + 4S_2 = 18$$
$$S_2 = 4\tfrac{1}{2}$$

These values are plotted on Fig. 6.2.
To verify that this is a straight line, other points can be plotted, e.g.

Let
$$S_2 = 1$$
$$8S_1 + 4 = 18$$
$$S_1 = \frac{18 - 4}{8} = 1\tfrac{3}{4}$$

Let
$$S_2 = 2$$
$$8S_1 + 8 = 18$$
$$S_1 = \frac{18 - 8}{8} = 1\tfrac{1}{4}$$

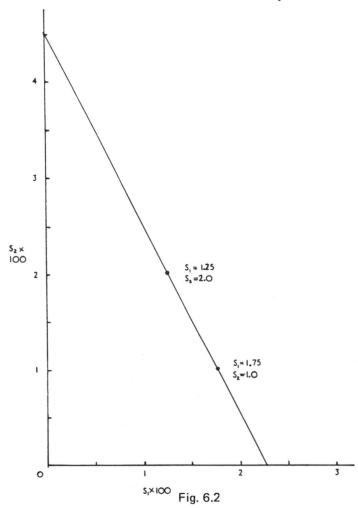

Fig. 6.2

The manpower requirements are now restricted to the area between this line and the origin since any point within the triangle or on the edge satisfies the constraint $8S_1 + 4S_2 \leqslant 18$ whilst any point outside fails to satisfy it.

If Figs. 6.1 and 6.2 are combined (see Fig. 6.3) it can be seen clearly that the area of choice is now limited to the area shown shaded. Any point outside this area will require either excessive manpower or greater production of slabs.

The objective function $8S_1 + 6S_2$ will now be considered. Take any particular profit margin and calculate what combination of S_1 and S_2 will give this profit; assume a profit of £8,

i.e. $$8S_1 + 6S_2 = £8$$

If $$S_2 = 0$$

$$8S_1 + 0 = £8$$

$$S_1 = 1$$

If $$S_1 = 0$$

$$0 + 6S_2 = £8$$

$$S_2 = 1\tfrac{1}{3}$$

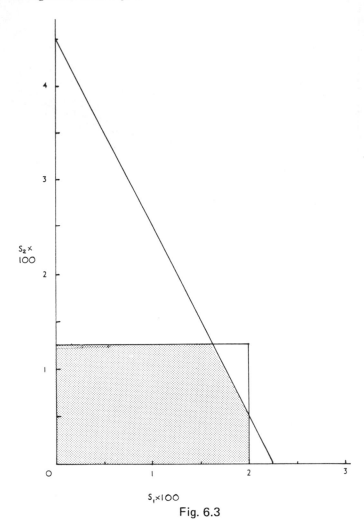

Fig. 6.3

This can be plotted on the graph (see Fig. 6.4).

For each point on the line the profit is £8 per day

e.g. $S_2 = \frac{1}{2}$

$$8S_1 + 3 = £8$$

$$S_1 = \frac{5}{8}$$

Now take a profit of £12 per day,

i.e. $8S_1 + 6S_2 = £12$

If $S_2 = 0$

$S_1 = 1\frac{1}{2}$

If $S_1 = 0$

$S_2 = 2$

This is also plotted on Fig. 6.4.

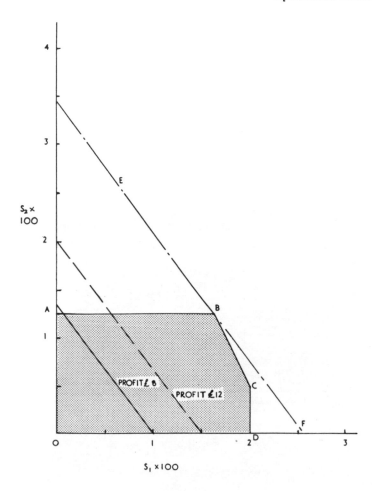

Fig. 6.4

It can be seen that the lines representing profits of £8 and £12 respectively are parallel and that the £12 profit line is further from the origin. These lines of variable profit may be likened to contour lines and the further these are from the origin the greater the profit margin will be. In this case it can be seen that the greatest profit margin allowed by the constraints lies on the line EF at the point B, i.e. $S_1 = 1\frac{5}{8}$, $S_2 = 1\frac{1}{4}$ this represents a profit of

$$8S_1 + 6S_2$$
$$= 8 \times 1\frac{5}{8} + 6 \times 1\frac{1}{4}$$
$$= 13 + 7\frac{1}{2}$$
$$= £20.50 \text{ per day}$$

by making $162\frac{1}{2}$ precast slabs and 125 hydraulically pressed slabs per day. Obviously it is not possible to make $162\frac{1}{2}$ slabs but this would represent 325 every two days.

This is the solution to this problem.

6.2.3.2 *Varying the profit margin*
Once the solution has been found for one profit margin it is very easy to provide a solution showing the effect of other profit margins.

Example 1
If the profit on precast slabs is to be £10 per 100, and on hydraulically pressed slabs is to be £5 per 100, the graphical solution will be as shown in Fig. 6.5.

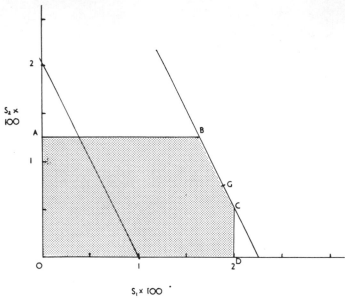

Fig. 6.5

The objective function would become

$$10S_1 + 5S_2$$

Take a particular profit and plot this on the graph. Assume a profit of £10

i.e. $$10S_1 + 5S_2 = £10$$

if $$S_1 = 0$$
$$S_2 = 2$$

if $$S_2 = 0$$
$$S_1 = 1$$

In this case the greatest profit margin can be achieved in a number of different ways, each one falling on the line BC

e.g.   at B $S_1 = 1\frac{5}{8}$ $S_2 = 1\frac{1}{4}$

$\therefore$ profit $= 10 \times 1\frac{5}{8} + 5 \times 1\frac{1}{4} = 22\frac{1}{2}$

at C $S_1 = 2$ $S_2 = \frac{1}{2}$

$\therefore$ profit $= 10 \times 2 + 5 \times \frac{1}{2} = 22\frac{1}{2}$

at G $S_1 = 1\frac{7}{8}$ $S_2 = \frac{3}{4}$

$\therefore$ profit $= 10 \times 1\frac{7}{8} + 5 \times \frac{3}{4} = 22\frac{1}{2}$

Example 2

If the profit on the precast slabs is to be £12.50 and on hydraulically pressed slabs £5 the graphical solution will be as shown in Fig. 6.6.

The objective function would be $12.5S_1 + 5S_2$.

Take a profit of £10.

i.e. $\qquad\qquad 12.50S_1 + 5S_2 = £10$

if $\qquad\qquad\qquad S_1 = 0 \quad S_2 = 2$

if $\qquad\qquad\qquad S_2 = 0 \quad S_1 = 0.8$

The greatest profit margin is now at point C and this represents a profit of

$$12.50S_1 + 5S_2$$
$$= 12.50 \times 2 + 5 \times \tfrac{1}{2} = £27.50$$

It can be seen that no matter what profit margin is used the solution always lies on the boundary of the feasable region and usually on a corner point. This is a fundamental property of linear programming formulations.

Fig. 6.6

6.2.4 The simplex technique

6.2.4.1 Introduction

As stated earlier, problems with more than two variables are not normally solved graphically and an algebraic solution is necessary. In order to solve the problem algebraically it is necessary to transform all the inequalities into equalities, e.g. $S_1 \leqslant 2$ is an inequality but by introducing S_3 to represent the excess of space on land bought by the builder this can be turned into an equality thus: $S_1 + S_3 = 2$.

6.2.4.2 Slack variables

The variable S_3 above is known as a SLACK VARIABLE and like other variables $S_3 \geqslant 0$. Where $S_3 = 0$ there is no excess of space on the land.

All the inequalities can be transformed into equalities by this method.

6.2.4.3 Paving slabs example

Referring to the example on the production of paving slabs, equations can be obtained in a similar fashion:

e.g. $S_1 \leqslant 2$ becomes $S_1 + S_3 = 2$

$S_2 \leqslant 1\frac{1}{4}$ becomes $S_2 + S_4 = 1\frac{1}{4}$ (S_4 represents unused capacity of the presses)

$8S_1 + 4S_2 \leqslant 18$ becomes $8S_1 + 4S_2 = S_5 = 18$ (S_5 represents unused labour)

It can be seen that there are three equations with five unknowns.

The solutions are represented by the corners in Fig. 6.4. At each of the corners two of the variables are zero, e.g. at point c the slack variable S_3 is zero as there is no space for storage of more precast slabs and the slack variable S_5 is zero because the labour force is fully utilised.

The solution normally lies on a corner point and at the corners two variables are zero and three are positive (in this case). The number of positive variables is normally the same as the number of side constraints. N.B. In some cases the slack variables may be positive and the decision variables zero.

As the solution lies on a corner point the aim is to find the corner points. Generally speaking, when there are n linear equations and n unknowns there is only one solution. There are exceptions to this rule but it is not necessary to consider the exceptions in this text.

The solution to this type of problem can be found by elimination but this method is very laborious and the *Simplex technique* is preferable in this situation.

6.2.4.4 Application of the simplex technique

Start at the origin when no precast or hydraulically pressed slabs are being produced (see Fig. 6.7).

1. $\qquad S_1 = 0, S_2 = 0, S_3 = 2, S_4 = 1\frac{1}{4}, S_5 = 18$

i.e. at origin 0 profit $= 8S_1 + 6S_2 = 0$.

Proceed by replacing zero variables (S_1 and S_2) with positive variables (S_3, S_4 and S_5). At each stage exchange one positive for one zero variable so that the objective function increases or remains as it was – if all possible switches decrease the value of the objective function then the solution has been found.

It is usual to increase the variable giving the largest increase in profit first, i.e. S_1.

2. $\qquad S_1 = 2, S_2 = 0, S_3 = 0, S_4 = 1\frac{1}{4}, S_5 = 2$

i.e. at point D profit $= 8S_1 + 6S_2 = £16$.

3. $\qquad S_1 = 2, S_2 = \frac{1}{2}, S_3 = 0, S_4 = \frac{3}{4}, S_5 = 0$

i.e. at point C profit $= 8S_1 + 6S_2 = £19$.

4. $\qquad S_1 = 1\frac{5}{8}, S_2 = 1\frac{1}{4}, S_3 = \frac{3}{8}, S_4 = 0, S_5 = 0$

i.e. at point B profit $= 8S_1 + 6S_2 = £20.50$

5. $\qquad S_1 = 0, S_2 = 1\frac{1}{4}, S_3 = 2, S_4 = 0, S_5 = 13$

i.e. at point A profit $= 8S_1 + 6S_2 = £7.50$.

Fig. 6.7

6.2.4.5 *Note*

Figure 6.7 is used simply to illustrate the movement from corner to corner. As stated earlier, in problems having a number of variables it would not be possible to illustrate the method using a diagram but the procedure used to calculate the optimum solution is as shown above.

It is simple to calculate the value of having one extra man in the gang by calculating the profit with one extra man and thus calculate the cost of employing him. This will show whether it is economical to do so.

Using the simplex technique it is possible to find the cost of restrictions as well as the solution to the problem. A linear programming problem is characterised by three requirements:—

1. There must be a function, linear in a number of decision variables which has to be maximised or minimised.

2. Decision variables must *not* assume negative values.

3. They must satisfy a number of linear side conditions, usually in the form of inequalities. Often the most difficult task is to recognise a problem as a linear programming problem.

6.3   THE TRANSPORTATION PROBLEM

6.3.1 Introduction

Transportation problems can be solved using the Simplex Technique but a better method has been developed. This method is sometimes called the 'Transportation Technique of Linear Programming'. It can be applied in situations where products or materials are produced in a number of different places for distribution to other places. The object is to optimise the effectiveness of the operation.

Basically the optimum solution is arrived at by progressively comparing the cost of routes initially selected with those not selected and systematically adjusting the schedule where costs are lower.

In general terms the objective is to allocate jobs to facilities in such a way as to optimise the effectiveness in terms of cost.

6.3.2 Example of transportation problem

A pre-mixed concrete firm has to supply concrete to three different projects, 1, 2, and 3. The projects require 200, 350 and 400 cubic metres of concrete respectively in a particular week. The firm has three plants, A, B, and C producing concrete and these can provide 250, 400 and 350 cubic metres respectively.

The problem is one of minimising costs, and the costs of transporting concrete from each plant to each project is given in Fig. 6.8.

TRANSPORT FROM MIXING PLANT	TRANSPORT TO PROJECT AT			CAPACITY OF MIXING PLANT
	1	2	3	
A	40	20	30	250
B	40	50	100	400
C	20	80	30	350
DAILY REQTS BY PROJECTS IN M³	200	350	400	

Fig. 6.8

The quantity of concrete to be supplied by each plant to each project must be determined.

Figure 6.8 is self explanatory. The figures in the body of the table represent the cost in new pence of transport from plant to project. E.g. It costs 80p per cubic metre to transport concrete from plant C to project 2.

The objective function which has to be minimised is $40x_{A_1} + 20x_{A_2} + 30x_{A_3} + 40x_{B_1} + 50x_{B_2} + 100x_{B_3} + 20x_{C_1} + 80x_{C_2} + 30x_{C_3}$. x represents the quantity of concrete to be sent from plant to project, i.e. x_{A_1} represents the quantity of concrete sent from plant A to project 1.

When all the costs are added, this will give the total cost of transportation.

The side conditions which must be imposed due to the limited capacity of each plant are

$$x_{A_1} + x_{A_2} + x_{A_3} \leqslant 250 \tag{1}$$

$$x_{B_1} + x_{B_2} + x_{B_3} \leqslant 400 \tag{2}$$

$$x_{C_1} + x_{C_2} + x_{C_3} \leqslant 350 \tag{3}$$

The side conditions which must be imposed by the requirements of each project are

$$x_{A_1} + x_{B_1} + x_{C_1} = 200 \tag{4}$$

$$x_{A_2} + x_{B_2} + x_{C_2} = 350 \tag{5}$$

$$x_{A_3} + x_{B_3} + x_{C_3} = 400 \tag{6}$$

And a side condition imposed by the excess capacity of the premixed concrete plant is

$$x_{A_4} + x_{B_4} + x_{C_4} = 50 \tag{7}$$

6.3.2.1 Slack variables

As the side conditions imposed by the limited capacity of each plant are inequalities, slack variables must be introduced. The side conditions can then be expressed as equations, thus:—

$$x_{A_1} + x_{A_2} + x_{A_3} + x_{A_4} = 250 \tag{8}$$

$$x_{B_1} + x_{B_2} + x_{B_3} + x_{B_4} = 400 \tag{9}$$

$$x_{C_1} + x_{C_2} + x_{C_3} + x_{C_4} = 350 \tag{10}$$

i.e. x_{A_4} represents the excess capacity of plant A
x_{B_4} represents the excess capacity of plant B
x_{C_4} represents the excess capacity of plant C

The decision variables must of course be non-negative.

The number of positive variables in a transportation problem is normally $m + n - 1$.

Where m is the number of rows and n is the number of columns — in this case $3 + 4 - 1 = 6$. This is because whilst there are 7 side conditions, one can easily be found if we know the remaining six, so there are in fact only six independent side conditions and therefore six decision variables, i.e. there are as many positive variables as there are side conditons.

6.3.2.2 *Finding an initial solution using the Northwest Corner Rule*

An initial solution must be found and this can be done using the table shown in Fig. 6.9 which has an additional column in which the excess capacity available at the mixing plants can be shown.

TRANSPORT FROM MIXING PLANT	TRANSPORT TO PROJECT AT			SURPLUS	CAPACITY OF MIXING PLANT
	1	2	3		
A					2 5 0
B					4 0 0
C					3 5 0
DAILY REQTS. BY PROJECTS IN M³	2 0 0	3 5 0	4 0 0	5 0	

Fig. 6.9

The selection of the decision variables in the body of the table can now be made using the Northwest Corner Rule, thus:—

1. Start at the northwest corner (x_{A_1} representing the quantity of concrete to go from plant A to project 1). Plant A produces 250 cubic metres, project 1 requires 200 cubic metres. Send the maximum amount to project 1, i.e. 200 cubic metres. This satisfies the requirements of project 1 and thus completes the first column (Fig. 6.10). Now move progressively right and down until x_{C_4} is reached thus.

TRANSPORT FROM MIXING PLANT	TRANSPORT TO PROJECT AT			SURPLUS	CAPACITY OF MIXING PLANT
	1	2	3		
A	2 00				2 5 0
B	0				4 0 0
C	0				3 5 0
DAILY REQTS BY PROJECTS IN M³	2 00	3 5 0	4 0 0	5 0	

Fig. 6.10

2. Plant A still has 50 cubic metres excess capacity. This is sent to project 2. This exhausts the supply from plant A and the first row is complete (Fig. 6.11).

TRANSPORT FROM MIXING PLANT	TRANSPORT TO PROJECT AT			SURPLUS	CAPACITY OF MIXING PLANT
	1	2	3		
A	2 OO	5 O	O	O	2 5 O
B	O				4 OO
C	O				3 5 O
DAILY REQTS BY PROJECTS IN M³	2 OO	3 5 O	4 OO	5 O	

Fig. 6.11

3. The new northwest corner is x_{B_2}. Project 2 needs 300 cubic metres more and this quantity is supplied from plant B. The second column is now complete (Fig. 6.12).

TRANSPORT FROM MIXING PLANT	TRANSPORT TO PROJECT AT			SURPLUS	CAPACITY OF MIXING PLANT
	1	2	3		
A	2 OO	5 O	O	O	2 5 O
B	O	3 OO			4 OO
C	O	O			3 5 O
DAILY REQTS BY PROJECTS IN M³	2 OO	3 5 O	4 OO	5 O	

Fig. 6.12

4. Plant B still has 100 cubic metres excess capacity. This is now sent to project 3 and second row is now complete (Fig. 6.13).

TRANSPORT FROM MIXING PLANT	TRANSPORT TO PROJECT AT			SURPLUS	CAPACITY OF MIXING PLANT
	1	2	3		
A	2 OO	5 O	O	O	2 5 O
B	O	3 OO	1 OO	O	4 OO
C	O	O			3 5 O
DAILY REQTS BY PROJECTS IN M³	2 OO	3 5 O	4 OO	5 O	

Fig. 6.13

5. Thè new northwest corner x_{C_3}. Project 3 needs 300 cubic metres more and this quantity is supplied from plant C (Fig. 6.14).

TRANSPORT FROM MIXING PLANT	TRANSPORT TO PROJECT AT			SURPLUS	CAPACITY OF MIXING PLANT
	1	2	3		
A	2 OO	5 O	O	O	2 5 O
B	O	3OO	I OO	O	4OO
C	O	O	3 O O		35 O
DAILY REQTS BY PROJECTS IN M³	2 OO	3 5 O	4 O O	5 O	

Fig. 6.14

6. Plant C still has 50 cubic meters excess capacity and this is shown in the surplus column (Fig. 6.15).

This gives a solution with exactly the right number of variables, which satisfies all the side conditions.

TRANSPORT FROM MIXING PLANT	TRANSPORT TO PROJECT AT			SURPLUS	CAPACITY OF MIXING PLANT
	1	2	3		
A	2 OO	5 O	O	O	2 5 O
B	O	3 OO	I OO	O	4 OO
C	O	O	3 OO	5 O	3 5 O
DAILY REQTS BY PPOJECTS IN M³	2 OO	3 5 O	4 O O	5 O	

Fig. 6.15

Notes

Whilst the Northwest Corner Rule gives an initial solution, it takes no account of cost. The total cost of transport in the above solution can be found thus:—

$$x_{A_1} = 200, x_{A_2} = 50, x_{B_2} = 300, x_{B_3} = 100,$$
$$x_{C_3} = 300, \text{ and } x_{C_4} = 50.$$

Therefore the total cost will be:

$$(200 \times 40) + (50 \times 20) + (300 \times 50) + (100 \times 100) + (300 \times 30) + (50 \times 0)$$

$$= 8000 + 1000 + 15000 + 10000 + 9000 + 0$$

$$= 43000p$$

$$= £430$$

6.3.2.3 Alternative method of finding an initial solution

It is in fact possible to develop an initial solution which is cheaper. To do this the table of costs (Fig. 6.8) is used and the procedure is as follows:—

1. Begin with the least cost. In this case two costs are the same, A_2 and C_1. Use A_2 and make this 250 which completes the first row (Fig. 6.16). Then continue by distributing concrete

TRANSPORT FROM MIXING PLANT	TRANSPORT TO PROJECT AT			SURPLUS	CAPACITY OF MIXING PLANT
	1	2	3		
A		250			250
B	200	100	100		400
C			300	50	350
DAILY REQTS BY PROJECTS IN M³	200	350	400	50	

Fig. 6.16

(a) to the site which received some but not all, concrete required at the end of the previous step or from the plant at which some remained at the end of the previous step.

(b) At least cost thus:

2. Complete second column by filling in x_{B_2} or x_{C_2}, choose the one which costs least, i.e. x_{B_2} and make this 100 which completes the second column.

3. Complete the second row by filling in x_{B_1} or x_{B_3}, choose the one with the least cost, i.e. x_{B_1} and make this 200 which is the total requirement for site 1. Now complete the second row by making x_{B_3} 100 and this exhausts the supply from plant B.

4. Complete the third column by filling in x_{C_1} or x_{C_3}. In this case the least cost is the same and we can therefore choose either. Choose x_{C_3} and make this 300 which is the total requirement of site 3.

5. Complete the third row by making x_{C_4} 50 which is the excess capacity of plant c.

The total cost in this case will be:

$$(250 \times 20) + (200 \times 40) + (100 \times 50) + (100 \times 100) + (300 \times 30) + (50 \times 0)$$

$$= 5000 + 8000 + 5000 + 10000 + 9000' = 0$$

$$= 37000p$$

$$= £370$$

this is a saving of £60.

6.3.2.4 The optimum solution to transportation problems

It is necessary to know if this solution is the optimum solution; and the following check will determine this. In a transportation problem the optimum solution remains so when all the results in a row or column are increased or decreased by the same amount, e.g. Using the table of costs (Fig. 6.8). Deduct 50p from all the costs in the second row and add 50p to all entries in the third column the result will be as shown in Fig. 6.17. The optimum solution for Fig. 6.8 and Fig. 6.17 is the same, i.e. in both cases it is necessary to send the same amount of concrete from each plant to each site. It is obviously impossible to have negative costs but it is convienient to use negative costs to determine whether a solution is optimum or not.

TRANSPORT FROM MIXING PLANT	TRANSPORT TO PROJECT AT			SURPLUS	CAPACITY OF MIXING PLANT
	1	2	3		
A	40	20	80	0	250
B	−10	0	100	0	400
C	20	80	80	0	350
DAILY REQTS BY PROJECTS IN M³	200	350	400	50	

Fig. 6.17

6.3.2.5 *Checking for optimum solutions*
To find whether the solution found is the optimum one the table of costs is adjusted so that the allocations used have a cost of zero. The signs of the costs not used are then considered. If the signs of costs not used are all positive, the solution is optimum. If any signs are negative the solution is not optimum. To do this calculation the table of costs Fig. 6.8 and the initial solution Fig. 6.16 are used and the procedure is as follows:—

1. Set out a table of costs for the distribution used, with an extra column and row (Fig. 6.18).

2. Make these costs zero by adding or deducting constant numbers. (This will not alter the optimum solution.)

	SITE 1 REQTS. 200	SITE 2 REQTS 350	SITE 3 REQTS 400	SURPLUS 50	
PLANT A CAPACITY 250		20			
PLANT B CAPACITY 400	40	50	100		40
PLANT C CAPACITY 350			30	0	
	0				

Fig. 6.18

(i) Start by inserting zero in the first column of the table of costs, Fig. 6.18.
(ii) Cost of 40p (B_1) must become zero, therefore deduct 40p from the second row, shown in last column.

(iii)   Cost of 50p (B_2) must also become zero, 40p is already shown in the second row, therefore deduct a further 10p, shown in second column. (40p + 10p) = 50p (Fig. 6.19).

(iv)   Similarly given 40p in the second row we need to deduct a further 60p shown in the third column (40p + 60p) = 100p (Fig. 6.19).

(v)   Proceed as above to give − 30p in the third row (60p − 30p) = 30p, 30p in the fourth column (30p − 30p) = 0, 10p in the first row (10p + 10p) = 20p (Fig. 6.19).

	SITE I REQTS. 200	SITE 2 REQTS. 350	SITE 3 REQTS. 400	SURPLUS 50	
PLANT A CAPACITY 250		20			10
PLANT B CAPACITY 400	40	50	100		40
PLANT C CAPACITY 350			30	0	− 30
	0	10	60	30	

Fig. 6.19

3. Return to the original table of costs and insert the numbers just calculated in the bottom row and last column (Fig. 6.20).

Deduct from all costs the cost in the bottom row and last column, e.g. cost 40p (A_1) becomes 40p − 10p − 0 = 30p. The complete table is shown in Fig. 6.21.

When all unused allocations are positive or zero the solution will be optimum. In this case there are negative costs and the costs can be further reduced.

The objective function is

$$30x_{A_1} + 0x_{A_2} - 40x_{A_3} - 40x_{A_4}$$

$$0x_{B_1} + 0x_{B_2} + 0x_{B_3} - 70x_{B_4}$$

$$50x_{C_1} + 110x_{C_2} + 0x_{C_3} + 0x_{C_4}.$$

	SITE I REQTS. 200	SITE 2 REQTS. 350	SITE 3 REQTS. 400	SURPLUS 50	
PLANT A CAPACITY 250	40	20	30	0	10
PLANT B CAPACITY 400	40	50	100	0	40
PLANT C CAPACITY 350	20	80	30	0	− 30
	0	10	60	30	

Fig. 6.20

	SITE 1 REQTS. 200	SITE 2 REQTS. 350	SITE 3 REQTS. 400	SURPLUS 50	
PLANT A CAPACITY 250	40-10-0 = 30	20-10-10 = 0	30-10-60 = -40	0-10-30 = -40	10
PLANT B CAPACITY 400	40-40-0 = 0	50-40-10 = 0	100-40-60 = 0	0-40-30 = -70	40
PLANT C CAPACITY 350	20-30-0 = 50	80--30-10 = 110	30-30-60 = 0	0-30-30 = 0	-30
	0	10	60	30	

Fig. 6.21

5.3.2.6 Finding the optimum solution

To find the optimum solution the value of the objective function must be decreased by making x_{A_3}, x_{A_4} or x_{B_4} positive. Choose x_{B_4} as it is multiplied by the largest negative number. This must be increased as much as possible taking account of the side constraints. Every increase decreases the value of the objective function. The procedure is as follows:—

(i) Using the last solution (Fig. 6.16) add a + sign to the zero of x_{B_4} (Fig. 6.22) as this will be increased to make it positive.

TRANSPORT FROM MIXING PLANT	TRANSPORT TO PROJECT AT			SURPLUS	CAPACITY OF MIXING PLANT
	1	2	3		
A	0	250	0	0	250
B	200	100	100⁻	0⁺	400
C	0	0	300⁺	50⁻	350
DAILY REQTS. BY PROJECTS IN M³	200	350	400	50	

Fig. 6.22

(ii) If concrete is retained at plant B (excess capacity) it will be necessary to send less to site 1, 2 or 3 (B_1, B_2 or B_3) and it will be necessary to retain less concrete at plant C. Attach a minus sign to C_4. If less concrete is retained at plant C more must be sent to site 3 (C_3). If more concrete is sent to site 3 from plant C (C_3) less will be required from plant B. This determines that it will be site 3 and not site 1 or 2 which receives less concrete from B.

(iii) x_{B_4} can be increased by a maximum of 50 cubic metres, if it was increased by more than this x_{C_4} would become negative. All allocations with a plus sign are then increased by 50 and those with a minus sign decreased by 50 (Fig. 6.23). The total cost is now:

(250 x 20) + (200 x 40) + (100 x 50) + (50 x 100) + (50 x 0) + (350 x 30)

= 5000 + 8000 + 5000 + 5000 + 0 + 10500

= 33500p

= £335.00

A further saving of £35.

TRANSPORT FROM MIXING PLANT	TRANSPORT TO PROJECT AT			SURPLUS	CAPACITY OF MIXING PLANT
	1	2	3		
A		2 5 0		0	2 5 0
B	2 0 0	1 0 0	5 0	5 0	4 0 0
C			3 5 0	0	3 5 0
DAILY REQTS. BY PROJECTS IN M²	2 0 0	3 5 0	4 0 0	5 0	

Fig. 6.23

To check whether this is the optimum solution the latest cost schedule is used (Fig. 6.24) and the same procedure is repeated (Figs. 6.24, 6.25 and 6.26).

This still leaves a negative cost and the solution is not therefore optimum. x_{A_3} must be increased and the procedure above (i) to (iii) is followed (Figs. 6.27 and 6.28).

The total cost is now

(200 x 20) + (50 x 30) + (200 x 40) + (150 x 50) + (50 x 0) + (350 x 30)

= 4000 + 1500 + 8000 + 7500 + 0 + 10500

= 31500p

= £315

A further saving of £20.

	SITE 1 REQTS. 200	SITE 2 REQTS. 350	SITE 3 REQTS. 400	SURPLUS 50	
PLANT A CAPACITY 250		2 0			1 0
PLANT B CAPACITY 400	4 0	5 0	1 0 0	0	4 0
PLANT C CAPACITY 350			3 0		− 3 0
	0	1 0	6 0	− 4 0	

Fig. 6.24

	SITE 1 REQTS. 200	SITE 2 REQTS. 350	SITE 3 REQTS. 400	SURPLUS 50	
PLANT A CAPACITY 250	40	20	30	0	10
PLANT B CAPACITY 400	40	50	100	0	40
PLANT C CAPACITY 350	20	80	30	0	-30
	0	10	60	-40	

Fig. 6.25

	SITE 1 REQTS. 200	SITE 2 REQTS. 350	SITE 3 REQTS. 400	SURPLUS 50	
PLANT A CAPACITY 250	30	0	-40	30	10
PLANT B CAPACITY 400	0	0	0	0	40
PLANT C CAPACITY 350	50	100	0	70	-30
	0	10	60	-40	

Fig. 6.26

TRANSPORT FROM MIXING PLANT	TRANSPORT TO PROJECT AT			SURPLUS	CAPACITY OF MIXING PLANT
	1	2	3		
A	0	250⁻	0⁺	0	250
B	200	100⁺	50⁻	50	400
C	0	0	350	0	350
DAILY REQTS. BY PROJECTS IN M³	200	350	400	50	

Fig. 6.27

TRANSPORT FROM MIXING PLANT	TRANSPORT TO PROJECT AT			SURPLUS	CAPACITY OF MIXING PLANT
	1	2	3		
A	0	200	50	0	250
B	200	150	0	50	400
C	0	0	350	0	350
DAILY REQTS BY PROJECTS IN M³	200	350	400	50	

Fig. 6.28

A check is again made to determine whether this is the optimum solution (Figs. 6.29, 6.30 and 6.31). This leaves no negative number and is therefore the optimum solution.

	SITE 1 REQTS. 200	SITE 2 REQTS. 350	SITE 3 REQTS. 400	SURPLUS 50	
PLANT A CAPACITY 250	—	20	30	—	10
PLANT B CAPACITY 400	40	50	—	0	40
PLANT C CAPACITY 350	—	—	30	—	10
	0	10	20	−40	

Fig. 6.29

	SITE 1 REQTS. 200	SITE 2 REQTS. 350	SITE 3 REQTS. 400	SURPLUS 50	
PLANT A CAPACITY 250	40	20	30	0	10
PLANT B CAPACITY 400	40	50	100	0	40
PLANT C CAPACITY 350	20	80	30	0	10
	0	10	20	−40	

Fig. 6.30

	SITE 1 REQTS. 200	SITE 2 REQTS. 350	SITE 3 REQTS. 400	SURPLUS 50	
PLANT A CAPACITY 250	30	0	0	30	10
PLANT B CAPACITY 400	0	0	40	0	40
PLANT C CAPACITY 350	10	60	0	30	10
	0	10	20	−40	

Fig. 6.31

6.4 THE ASSIGNMENT PROBLEM

6.4.1 Introduction

Like transportation problems, assignment problems can be solved using the Simplex Technique, but again a better method has been developed.

This technique can be applied in situations where a fixed number of operatives are available to do a fixed number of jobs or operations. The object is to allocate jobs to facilities in such a way as to optimise the overall effectiveness.

6.4.2 Example in the construction industry

Assume the majority of a contractor's work is local authority housing and he wishes to expand and carry out larger housing projects. He has decided to use the Line of Balance method of planning and his carpenters will therefore be specialising in a limited area of work. At present his carpenter gangs each work on the whole range of operations, and on one project, the contractor has analysed the feed back data relating to these gangs as shown in Fig. 6.32. The problem is to decide which gangs should specialise in the various operations.

6.4.3 Procedure for solving problem

1. Select the operation which takes the minimum time to perform in each column and fill in the time required for this operation beneath the column (see Fig. 6.32), e.g. in column 4 first fix takes the least time to perform at 33 hours. Fill in 33 beneath column 4.

Operation	Gang No.	1	2	3	4	5	6
Carcassing (floors and roofs)		110	100	95	105	115	120
First fix		36	30	40	33	42	32
Second fix		112	100	105	115	120	85
Finishings		30	25	40	35	35	28
Floorboards and staircase		40	37	45	48	50	35
First floor partitions		90	105	90	78	85	100
		30	25	40	33	35	28

Fig. 6.32

2. Subtract the minimum time in each column from all the other operation times (see Fig. 6.33). E.g. in column 1 carcassing takes 110 hours. Subtract 30 hours and this leaves 80 hours.

3. Select the minimum time in each row and fill in the time at the end of the row (see Fig. 6.33), e.g. in the first row 55 is the minimum operation time, fill in 55 at end of row.

Operation	Gang No.	1	2	3	4	5	6	
Carcassing (floors and roof)		80	75	55	72	80	92	55
First fix		6	5	0	0	7	4	0
Second fix		82	75	65	82	85	57	57
Finishings		0	0	0	2	0	0	0
Floorboards and staircase		10	12	5	15	15	7	5
First floor partitions		60	80	50	45	50	72	45

Fig. 6.33

4. Subtract the minimum time in each row from all the other times (see Fig. 6.34).

5. Draw vertical and horizontal lines through the rows and columns containing zeros, starting with those with most zeros. A solution has been found when a total of six lines are required to cover all the zeros. In Fig. 6.34 there are only four lines and a solution therefore has not been found.

6. Select the smallest number not covered by a line, i.e. five, and subtract this number from all numbers not covered. Where numbers are covered twice, i.e. at intersections, add 5 (see Fig. 6.35).

Operation	Gang No.	1	2	3	4	5	6
Carcassing (floors and roof)		25	20	0	17	25	37
First fix		6	5	0	0	7	4
Second fix		25	18	8	25	28	0
Finishings		0	0	0	2	0	0
Floorboards and staircase		5	7	0	10	10	2
First floor partitions		15	35	5	0	5	20

Fig. 6.34

Operation	Gang No.	1	2	3	4	5	6
Carcassing (floors and roof)		20	15	0	17	20	37
First fix		1	0	0	0	2	4
Second fix		20	13	8	25	23	0
Finishings		0	0	5	7	0	5
Floorboards and staircase		0	2	0	10	5	2
First floor partitions		10	30	5	0	0	27

Fig. 6.35

7. Repeat stage 5 as shown in Fig. 6.35. This gives six lines which will result in a solution. To find a solution, proceed as follows.

(a) Find a row with a single zero and assign that operation to the gang in that column. E.g. first row has one zero under gang 3 therefore carcassing is done by gang 3.

(b) Cancel all zeros in gang 3 column.

(c) Find next row with a single zero and assign as in (a) above, e.g. third row has one zero under gang 6, therefore second fix is done by gang 6.

(d) Cancel all zeros under gang 6.

(e) Proceed as (c), thus assigning floor boards and staircase to gang 1.

(f) No other rows have a single zero. Inspect columns for a single zero. In this case all remaining rows and columns have more than one zero. There will therefore be more than one solution. Allocate finishings to gang 2.

(g) Cancel all zeros in gang 2 column.

(h) Proceed as above allocating First Fix to gang 4 and First Floor Partitions to gang 5. This gives assignments as follows:

Gang 1 will specialise in Floor Boards and Staircase
Gang 2 will specialise in Finishings
Gang 3 will specialise in Carcassing
Gang 4 will specialise in First Fix
Gang 5 will specialise in First Floor Partitions
Gang 6 will specialise in Second Fix

The total work content of this solution is

$$40 + 25 + 95 + 33 + 85 + 85 = 363 \text{ hours.}$$

An alternative solution could be

Gang 1 will specialise in Floor Boards and Staircase
Gang 2 will specialise in First Fix
Gang 3 will specialise in Carcassing
Gang 4 will specialise in First Floor Partitions
Gang 5 will specialise in Finishings
Gang 6 will specialise in Second Fix

This also gives a work content of

$$40 + 30 + 95 + 78 + 35 + 85 = 363 \text{ hours}$$

It may not always be possible to use the gangs in accordance with these solutions the whole of the time, but wherever possible the allocations shown should be used.

6.5 QUEUING THEORY

6.5.1 Introduction

Queuing situations occur every day in one form or another. There are queues for buses, queues in the supermarket and other less obvious queues such as callers queuing for a vacant line on telephone switchboards and components queuing up waiting to be assembled in a factory. In fact in a factory, queues of partly assembled products are there by design to act as a buffer stock and thereby ensure continuity of work for operatives. This type of queue also occurs in the production of repetitive units in the construction industry and can be seen in the form of activity buffers, and stage buffers in the Line of Balance Method (see Fig. 1.32).

In some cases however queues of work waiting to be done can be very expensive. An example here could be repairs to plant, as this could cause hold-ups to other operatives.

In repetitive work such as housing, it is important to realise the effects of queues and their influence on production planning and control.

All queuing problems have one thing in common, a service and a customer, i.e. someone or something waiting to be served, e.g. houses waiting to be tiled by a roofer, excavators waiting to be repaired by mechanics, etc. The problem is usually one of economics, to compare the cost of providing an immediate service with the cost of keeping the 'customer' waiting.

Queuing Theory is based on probability theory because customers usually arrive at random intervals and usually take differing lengths of time to serve.

6.5.2 Queuing models

Mathematical models can be used to represent the real life situations and calculations made to find the likely waiting time in a queue, and the effect on this waiting time and costs if the service is improved, or reduced. It is important to consider other possible effects such as loss of customers due to excessive waiting, etc., where this is relevant.

6.5.3 Traffic intensity

Traffic Intensity or *Utilisation factor* is the ratio of mean overall service time to mean interarrival time (mean interval between arrivals).

i.e.   Traffic Intensity $(\rho) = \dfrac{\text{mean service time}}{\text{mean interarrival time}}$ or $\dfrac{S}{A}$

6.5.4 Mean waiting time

In queuing problems the mean waiting time is often required and this is found as follows:—

Mean waiting time including time being served (total time in the system)

$$= \frac{1}{1 - \rho} \times S = \frac{1}{1 - \dfrac{S}{A}} \times S = \frac{S}{\dfrac{A - S}{A}} = \frac{AS}{A - S}$$

If time being served is to be excluded, the formula is

$$\frac{AS}{A - S} - S = \frac{AS - S(A - S)}{A - S} = \frac{AS - AS - S^2}{A - S}$$

$$= \frac{S^2}{A - S}.$$

These formulae apply to situations where there is one service point and only apply after the system has become stabilised. This depends on the traffic intensity which must be less than 1. The nearer traffic intensity is to 1, then the longer it takes for the system to become stabilised. The formulae also apply strictly for random arrivals and service times which have a negative exponential distribution.

As the traffic intensity approaches 1 the mean waiting time increases rapidly and hold-up results. Obviously with traffic intensities greater than one, the queue will grow indefinitely. If however low traffic intensities are used in order to meet demand immediately it occurs, high nonproductive time results for the server. A traffic intensity of 0.6 means that the server is busy for only 60% of the time on average.

It is possible to have higher traffic intensities for normal periods wherever extra help can be brought in or overtime can be worked at busy periods. If this is not possible, it is better to have traffic intensities of say 0.7 to 0.8.

Waiting time can be reduced if arrivals can be made more regular and/or service times can be made approximately equal, i.e. a small standard deviation.

If arrivals come at exactly equal intervals and service times were constant, there would of course be no queue even with traffic intensities of 1.

Conversely, the more irregular the arrivals and service times the longer the waiting times will be with a given traffic intensity.

6.5.5 Example in the construction industry

A planning engineer in a construction firm is specialising in pre-tender planning. He has been working excessive overtime and still cannot completely meet the demand. The problem is to decide the best course of action in these circumstances. Information has been collected to obtain the average time necessary for a planning engineer to complete his part of a pre-tender plan and this was found to be 21 hours or 3 normal working days. The time normally available for him to complete his part of the plan is 9 working days from the time he receives information from the estimator. Information was also collected of the period of time elapsing between receipts of successive requests for pre-tender planning. The results are shown in Fig. 6.36.

6.5.6   Solution

6.5.6.1 *Traffic intensity*

$$\text{Average service time} = 3 \text{ days}$$

$$\text{Average inter-arrival time} = 4 \text{ days}$$

$$\therefore \text{ Traffic intensity} = \tfrac{3}{4} \quad = 0.75$$

∴ On average the planner is busy doing planning only 75% of his normal time.

6.5.6.2 *Mean waiting time*

The time available from receipt to completion of the work is 9 days. This includes time when the work is actually being done. The formula to use is therefore the one which includes time in the system

i.e. $$\text{Mean waiting time} = \frac{AS}{A - S} = \frac{4 \times 3}{1} = 12 \text{ days}$$

This indicates why the planning engineer has been working so much overtime.

If two planners are used assisting each other and tackling one project at a time, the traffic intensity would be reduced to $\dfrac{1\frac{1}{2}}{4} = 0.375$ and the waiting time would be reduced to $\dfrac{4 \times 1\frac{1}{2}}{2\frac{1}{2}} = 2.4$ days.

This would of course leave both planning engineers unoccupied for a large proportion of their time.

Arrival Number	Inter Arrival Time (days)	Arrival Times (days)	Cumulative Average Inter Arrival Time (days)
1	5	5	5.0
2	6	11	5.5
3	6	17	6.3
4	4	21	5.3
5	3	24	4.8
6	3	27	4.5
7	4	31	4.4
8	2	33	4.1
9	4	37	4.1
10	5	42	4.2
11	6	48	4.4
12	3	51	4.3
13	2	53	4.1
14	1	54	3.9
15	5	59	4.0
16	6	65	4.1
17	4	69	4.1
18	2	71	4.0
19	3	74	3.9
20	5	79	4.0
21	4	83	4.0
22	6	89	4.0
23	3	92	4.0
24	4	96	4.0
25	3	99	4.0
26	7	106	4.1
27	5	111	4.1
28	2	113	4.0
29	2	115	4.0
30	4	119	4.0
31	3	122	3.9
32	6	128	4.0
33	5	133	4.0

Fig. 6.36

6.5.6.3 A compromise solution

A planning engineer who normally specialises in planning after award of the contract could be trained in pre-tender planning and could then assist when necessary.

Figures 6.37 and 6.38 show the effect of traffic intensity on waiting time, assuming inter-arrival time will be 4 days.

6.5.6.4 Assumptions made

It is assumed that the pre-tender planning is completed on a first come, first served basis and that when assistance is required both planners work on the same project.

Inter Arrival Time (days)	Traffic Intensity	Service Time (days)	Waiting Time $\dfrac{AS}{A-S}$
4	0.1	0.4	0.44
	0.2	0.8	1.0
	0.3	1.2	1.71
	0.4	1.6	2.67
	0.5	2.0	4.0
	0.6	2.4	6.0
	0.7	2.8	9.33
	0.8	3.2	16.00
	0.9	3.6	36

Fig. 6.37

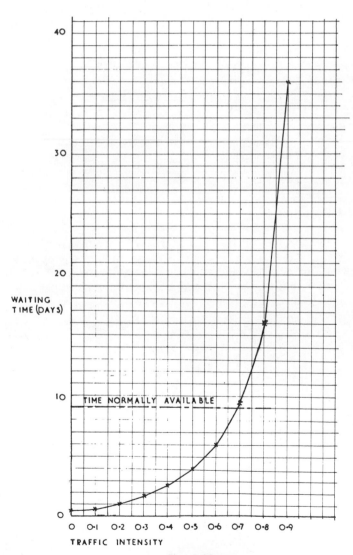

Fig. 6.38

6.5.7 Other examples in the construction industry

Many examples of queuing occur in the construction industry, e.g.

   (i)   Estimates queuing to be prepared in overworked estimating departments or estimaters idle because few estimates are arriving.

  (ii)   Lorries queuing up to be filled by excavating machines or excavators standing idle awaiting lorries for disposal of excavated material.

 (iii)   Concrete gangs idle awaiting completion of formwork and reinforcement.

 (iv)   Precast concrete panels waiting to be moved by cranes.

  (v)   The various stages of repetitive housing waiting for operatives to proceed with further operations.

This last example deserves particular attention because a considerable proportion of the total building effort is in housing of various types.

A considerable amount of work has been done on repetitive housing by the Building Research Station and the following is based on their work.

The mean service time is the average time taken to perform a particular operation on a unit. The mean arrival time is the average of the intervals between the completions of the previous unit. Obviously if mean service time is longer than mean arrival time then a queue of houses will result between the two operations and this queue will grow continuously.

If the mean arrival time is longer than the mean service time, operatives will be idle at times. If traffic intensity is one, i.e. equivalent to balanced gangs and no buffers are provided, variations in productivity, inaccuracies in targets set, poor materials, etc., will result in times when a queue of partially completed houses builds up and times when no houses are ready to be 'served'.

This can be overcome to some degree by allowing a buffer between operations or stages as shown in Fig. 1.26 and Fig. 1.32, i.e. deliberately form a queue. This will of course increase the time of production of a house, resulting in a greater capital investment in many cases.

In practice, the queuing problem becomes very complex, as there may be two gangs engaged on the same operation (see Fig. 6.39), operatives may do more than one job on a particular house (see Fig. 6.40) or two or more operations can be done on a house at one time, which gives a result similar to Fig. 6.39. In these circumstances, a mathematical model becomes very difficult and sometimes impossible to develop and simulation is more useful.

Fig. 6.39

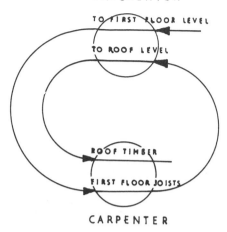

BRICKLAYER

TO FIRST FLOOR LEVEL

TO ROOF LEVEL

ROOF TIMBER

FIRST FLOOR JOISTS

CARPENTER

Fig. 6.40

6.6 SIMULATION

6.6.1 Introduction

As stated earlier some problems are too complex to solve mathematically and simulation can often be used in these cases. It can be used to find bottlenecks, labour and plant utilisation, throughput time, etc.

To help solve problems by simulation a model is used which represents the real system as nearly as possible.

A mechanism is necessary to simulate what is likely to happen in practice, i.e. to enable operation under working conditions. This may be discs numbered in accordance with the expected distribution, tables of random numbers, a die, or a computer programme.

Simulation is also used to verify the results of the application of analytical techniques.

6.6.2 Advantages of simulation

One advantage of simulation is that it does not become much more difficult as the problem becomes more complex. It is these complexities which often make analytical techniques impossible.

6.6.3 Simulation in construction work

When planning construction work the operation times should be calculated from output data from previous projects or from work study synthetics. In practice these operation times will vary and progress will not be as planned for a number of reasons, e.g. inaccurate operation times, variability in productivity of the operatives, bad weather, drawings, etc. not arriving on time and other factors out of the control of the operatives.

By studying the variability of output from expected durations of operations it is possible to simulate operation times and thus study what is likely to happen under differing conditions. Different methods of control can be compared and decisions made as to which method is likely to be most effective. Obviously this is a much cheaper way of assessing results than trying them out in practice.

6.6.4 Example of the use of simulation

As an example of the use of simulation a project will be considered consisting of 10 identical units each unit having 5 operations, with separate gangs on each operation as shown in Fig. 6.41 (repeated from page 41).

Operation	Number of Men to be used	Time required
Excavation	2	2
Concrete foundations	4	4
Brickwork	8	4
Roof construction	3	4
Roof finishing	2	4

Fig. 6.41

The project is planned using the Line of Balance Method with buffers, i.e. the operations will be planned to form a series of queues (see Fig. 6.42)

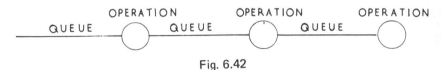

Fig. 6.42

From the Line of Balance Schedule in Fig. 6.43 it can be seen that the project should be completed in 66 days.

Progress can now be simulated using cards numbered to represent the time necessary to complete the respective operations, the numbers having a statistical distribution similar to that found in practice. The Line of Balance Schedule represents the model and the numbered cards are the mechanism.

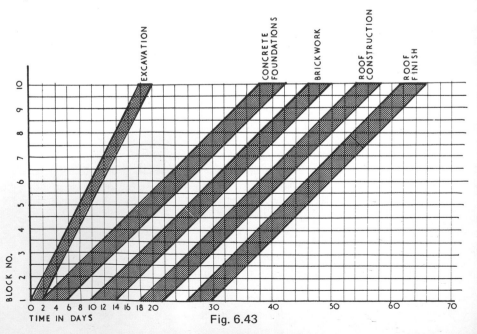

Fig. 6.43

6.6.4.1 Simulating progress
Progress is simulated for each operation on each block and the results are shown
on Fig. 6.44. The results are entered onto the Line of Balance Schedule
(Fig. 6.45).

Operation	Block No.	1	2	3	4	5	6	7	8	9	10
EXCAVATION											
start		0	2.0	3.6	5.4	7.8	10.2	12.4	14.6	16.4	18.2
simulated operation time		2.0	1.6	1.8	2.4	2.4	2.2	2.2	1.8	1.8	1.6
finish		2.0	3.6	5.4	7.8	10.2	12.4	14.6	16.4	18.2	19.8
CONCRETE FOUNDATIONS											
start		2.0	6.4	10.8	15.4	18.6	22.6	27.4	32.6	36.2	
simulated operation time		4.4	4.4	4.6	3.2	4.0	4.8	5.2	3.6		
finish		6.4	10.8	15.4	18.6	22.6	27.4	32.6	36.2		
BRICKWORK											
start		10.0	14.6	19.8	24.8	30.8	36.6				
simulated operation time		4.6	5.2	5.0	6.0	5.8					
finish		14.6	19.8	24.8	30.8	36.6					
ROOF CARCASS											
start		18.0	21.4	25.2	30.8	36.6					
simulated operation time		3.4	3.8	4.0	4.2						
finish		21.4	25.2	29.2	35.0						
ROOF FINISH											
start		26.0	30.2	35.0							
simulated operation time		4.2	4.8								
finish		30.2	35.0								

Fig. 6.44

Fig. 6.45

In this particular simulation the results show that progress was as planned until the roof carcass gang want to move into block 4 and they were delayed by 1.6 days. From that time until day 38 (the time up to which progress is shown on the chart), the roof carcass gang is being delayed. This would result in non-productive time and would therefore be costing extra money, unless of course the gangs are adjusted.

6.6.4.2 *Comparisons with other schedules*
The total cost of the project including overheads and all other factors can be ascertained on completion and a comparison can be made with the planned cost. Comparisons can also be made with a schedule using different buffer times between activities or by using different methods of control, e.g. progress charts in bar chart form or decision rules.

6.6.4.3 *The time factor*
The simulation of each method would have to be carried out a number of times to get a range of likely results and this would be very time consuming. Use of a computer would cut the time factor quite considerably and allow comparisons to be made more quickly.

6.6.4.4 *Effect of buffer on direct costs*
If no buffer had been provided between activities, the direct costs would have been much higher than the original schedule because the variability of actual and scheduled times would cause queues of blocks to build up in some cases and operatives to be non-productive in other cases. This is to be expected from what has been discussed in the section on Queuing Theory.

6.6.5 Conclusion
As stated earlier, simulation is extremely useful for helping to solve Management problems. It has a very wide range of application and whilst it has already been used in the Construction Industry, there is scope for extension of the use of this technique.

6.7   MANAGEMENT GAMES

6.7.1 Introduction
In management games the participants assume the various management roles and have to make decisions. An analysis of the decisions taken can be made on completion to assess the performance of those taking part. The games simulate real life situations as nearly as possible and rules are imposed to keep the games within acceptable limits.

6.7.2 Example of a management game
The simulation example described in 6.5.4 could be used as a basis for a management game.

6.7.2.1 *The estimate*
An estimate could be prepared for the project to include the usual costs met up with in practice, e.g. labour, plant, materials, and overheads.

6.7.2.2 *Planning and controlling*
The participants can then plan the project and by feeding information to them to represent weekly or daily progress, they up-date their own schedule.

6.7.2.3 Varying operation times
Provision can be made to vary output of operatives by allowing use of overtime or by increasing or decreasing gang sizes as would be possible in practice. This would of course add to the cost of the project.

6.7.2.4 Assessment of results
The results achieved by the participants could be compared with the estimate at intervals and on completion. Comparisons can also be made with other teams playing the game.

6.8 DECISION RULES

6.8.1 Introduction

Decision rules are based on queuing theory. In order to cut down the average waiting time in practice priority can be given to some "customers", this is known as varying the queue discipline. The remainder of this section on decision rules is taken from 'The Control of Repetitive Construction' by J. F. Nuttall of the Building Research Station. The work done by BRS included simulation exercises but the rules were not tested in practice.

6.8.2 Dynamic control

For a method of control to be dynamic, decisions need to be based on the latest available information. This information may be obtained from a programme adapted by the site manager as necessary. This adaptation of the programme can be formalised which can be very useful to the less experienced site manager. A simple example of this formalisation is to treat the sequence of operations in the programme as the order of priority for the operations.

6.8.3 Decisions based on experience

There are many contracts where no detailed programme exists and the site manager decides from day to day what work is to be done. Thus the decisions on sequence are based on the situation which exists at the moment of decision, as interpreted by the site manager's intuition and experience. This gives a fully dynamic control method which can give good results provided that the site manager is experienced in the type of work and is able to give due consideration to the various factors which should influence his decision.

6.8.4 Formal decision rules

Formal dynamic decision rules do exist which have advantages in being consistent, logical, objective and simple to apply and thus may be of great assistance to a site manager, particularly if he is not very experienced or if the type of construction is complex or unfamilar.

A number of decision rules were tested by the Building Research Station by simulation for a contract for semi-detached houses to give a preliminary assessment of the success they would be likely to have if applied in practice. These rules were used every time a trade was ready to start a fresh operation. The rules took into consideration all the operations which were available to be started at that time and assigned to each of them a certain priority. The operation with the highest priority was then chosen to be performed next.

6.8.4.1 Simple decision rules

The first decision rule to be tested was very simple: it was to choose the operation which would take least time to perform. This gave a fixed order of priority to the operations, e.g. the choice for the bricklayers would always be to do the topping out of any block which had reached that stage, or failing that, to build the ground floor brickwork of the next block due to be started, and the first floor brickwork would only be constructed when there was no other work. However, as in all the decisions rules tested, operations were always carried through to their completion without interruption. It was necessary, with this decision rule, to regard the blocks as becoming available for the first operation at intervals. The second decision rule was also very simple: in it, the priority of the operations depended entirely on the length of time for which they had been available to be performed Thus, in the queue of houses waiting to be operated on by any particular trade, the men would go first to the house which had been waiting longest since it was vacated by the previous trade. In applying this 'first come, first served' rule it was necessary to prevent the first operation in the houses from always having priority (since a house on which no work had been done would inevitably have been waiting longer than a partly built house). So the houses were released for the first operation periodically, as for the first decision rule.

The third decision rule was partly based on the first. This rule involved looking ahead to the trade which would follow the operation under consideration and giving priority to the trade which had least work ahead of it. When this did not fully determine the operation to be performed, the shortest operation rule was used. Thus in choosing an operation for the bricklayers, if there was less work ready and waiting for the tilers to do than there was for the carpenters, priority was given to topping out so as to release work for the tilers. If there was less work waiting for the carpenters than for the tilers, however, priority was given to ground floor brickwork or, failing that, to first floor brickwork, either of which would release work for the carpenters, ground floor brickwork being the first choice because it was the shorter operation.

6.8.4.2 Decision rules based on the estimated finishing times of the operations

The remaining decision rules which were tested were based on rules suggested by Rowe and Jackson. Each operation was given a priority based on a simple formula which, in effect, gave an approximate estimate of the day on which the operation concerned should be finished.

Suppose that the blocks are numbered $1, 2, 3, \ldots$ in the order in which they are to be constructed and that the operations are numbered $1, 2, 3, \ldots$ in the order in which they must be performed. Suppose further that

S_b = starting date of block b

d_i = estimated duration of operation i in working days per block

c = cycle time (i.e. maximum number of days' work per block for any one gang).

Consider a certain operation, a, on block b and estimate when it can be expected to be finished.

The successive blocks will reach any specified stage of construction at intervals of approximately c working days: in particular, their starting dates will be about c days apart. If dates are measured in working days from the commencement of the first block, which is therefore on day 0, the starting date of block b can be estimated approximately as:

$$S_b = c\,(b - 1) \tag{1}$$

It would be possible to perform all the operations on block b up to and including operation a within a period of:

$$\sum_{i=1}^{a} d_i \text{ working days}$$

$$\left(\begin{array}{l} \qquad \text{denotes the sum of the estimated durations} \\ \sum_{i=1}^{a} d_i \quad \text{of all the operations up to and including} \\ \qquad \text{operation } a \text{ in working days per block.} \end{array} \right)$$

However, this assumes that work proceeds continuously on the block without any breaks, whereas in practice there will be periods when the block is waiting between operations because the trade which is to perform the next operation is occupied elsewhere. A simple way of allowing for this is to increase the time allowed for the operations up to and including operation a by a factor f which is greater than unity, i.e. to allow a period of:

$$f \sum_{i=1}^{a} d_i \text{ working days}$$

for operations $1, 2, \ldots \ldots, a$ on any block. An approximate estimate of the day on which operation a on block b should be finished is therefore:—

$$P_{ab} = S_b + f \sum_{i=1}^{a} d_i \tag{2}$$

where S_b is obtained from formula (1).

Since the numbers P_{ab} approximately represent the dates by which the various operations should be completed, it is reasonable to use them as priority numbers for a decision rule, priority being given to the operation with the lowest priority number, i.e. the earliest estimated finishing date.

It might be thought better to take the estimated *starting* dates of the operations as their priority numbers, but the finishing dates have the advantage of giving a certain amount of priority to operations which can be performed quickly and will thus release work more quickly for the following trades.

Formula (2) gives a series of different decision rules, depending on the value which is taken for the factor f. In the simulation studies a number of these rules were tested, with values of f between one and three, so as to find the rule which gave the best results for the two-storey housing contracts. The priority numbers for the rule with $f = 2$ are given in Fig. 6.46 for the first four blocks to be built.

Priority numbers for decision rule with $f = 2$

Operation	Op. no. (a)	Trade initial	Estimated duration (d_a days)	fd_a	Block no. (b) 1	2	3	4
Approx. starting date (S_b)					0	$5\frac{1}{2}$	11	$16\frac{1}{2}$
Ground floor brickwork	1	B	$2\frac{1}{4}$	$4\frac{1}{2}$	$4\frac{1}{2}$	10	$15\frac{1}{2}$	21
First floor joists	2	C	$1\frac{1}{2}$	3	$7\frac{1}{2}$	13	$18\frac{1}{2}$	24
First floor brickwork	3	B	$2\frac{1}{2}$	5	$12\frac{1}{2}$	18	$23\frac{1}{2}$	29
Roof timbers	4	C	$2\frac{1}{2}$	5	$17\frac{1}{2}$	23	$28\frac{1}{2}$	34
Topping out	5	B	$\frac{3}{4}$	$1\frac{1}{2}$	19	$24\frac{1}{2}$	30	$35\frac{1}{2}$
Felt, battens & tiles	6	T	5	10	29	$34\frac{1}{2}$	40	$45\frac{1}{2}$
Staircase	7	C	$1\frac{1}{2}$	3	32	$37\frac{1}{2}$	43	$48\frac{1}{2}$

Fig. 6.46

The information required for obtaining the priority numbers consisted of the first four columns of the table, i.e. a list of the operations in the order in which they were to be performed, with their estimated durations in working days per block for the groups of men who were to perform them. The trade strengths had been chosen so that they were balanced as nearly as possible; thus the brick-layers had a total of $5\frac{1}{2}$ days' work per block, the carpenters $5\frac{1}{2}$ and the tilers 5 days' work. The estimated cycle was therefore $5\frac{1}{2}$ working days per block.

As an alternative to calculating each priority number individually from formula (2) the following procedure could be adopted. First, fill in the starting dates of the blocks, commencing with block 1, which is to start on day 0. The approximate starting dates of the following blocks are obtained in succession by adding the cycle time of $5\frac{1}{2}$ days to the starting date of the previous block.

Next, multiply each estimated duration by the factor f, in this case 2, to obtain the numbers in the column headed 'fd_a'. Now calculate the priority numbers, working down each column in turn and adding the corresponding value of fd_a to the number immediately above the number being calculated. For example, for the column headed 'Block 2' the successive calculations are: $5\frac{1}{2} + 4\frac{1}{2} = 10$, $10 + 3 = 13$, $13 + 5 = 18$, $18 + 5 = 23$ and so on.

The table of priorities can be calculated before the contract begins and only needs revising if the estimates of the operation durations change, e.g. if the labour available varies from the planned labour strengths.

To use the table in controlling the site work is very simple. Whenever a trade becomes free to undertake a fresh item of work and there is more than one item available for them to do, the item with the smallest priority number in Fig. 6.46 is chosen to be done first. For example, if the bricklayers have just finished the ground floor brickwork of block 3 and have the choice between commencing the ground floor brickwork of block 4, the first floor brickwork of block 2 or topping out on block 1, all of which operations are ready to be started, the decision rule states that the first floor brickwork of block 2 should be done, because this operation has a priority number of 18 compared with 19 and 21 for the other possibilities.

6.8.5 Random decisions

A small-scale simulation was also performed to assess the results of determining the sequence of operations by a purely random choice between the operations available at the time. This simulation gave a datum point with which to compare the various control methods tested, the differences between the results for the control methods and the results for the random decision giving a measure of the savings which can be achieved by good control compared with no control of the sequence of operations for this type of contract. An upper limit to the average savings obtainable was provided by the times given in the original programme, which represented an optimum or near-optimum solution for constant operation durations, but in practice this limit would never be attainable because of the impossibility of predicting the future accurately when operation durations vary from block to block.

6.8.6 Results of simulations

The simulations of dynamic control methods were based on the contract for semi-detached houses which had been used for earlier simulations, under the conditions of variation in operation times from block to block. The results are given in Fig. 6.47 and illustrated in Fig. 6.48.

Comparison between alternative methods of control

Methods of control	No. of contracts simulated	Contract duration (days)		Unproductive time (man-days)	
		Mean	Range	Mean	Range
Random decision	2	103.3	$93\frac{3}{4}-112\frac{3}{4}$	97	79–115
Programme (strict)	18	82.3	77–89	62	22–93
Programme (adapted)	10	81.6	$77\frac{1}{2}-87\frac{1}{2}$	45	23–69
Decision rules:					
Shortest operation 1st	4	87.3	$85-90\frac{1}{2}$	75	50–91
First come, first served	5	84.4	$81\frac{3}{4}-88\frac{1}{2}$	71	56–110
Look ahead to following trade, otherwise shortest operation 1st	4	82.3	$76\frac{1}{4}-93$	52	32–85
Formula with $f = 2$	14	81.7	$75\frac{3}{4}-84\frac{1}{2}$	47	22–76
$f = 1.7$	5	80.8	$76\frac{1}{2}-83$	43	38–51
$f = 1.6$	4	82.7	$80\frac{1}{4}-86$	46	37–56
Times shown in programme	—	$78\frac{1}{2}$		29	

Fig. 6.47

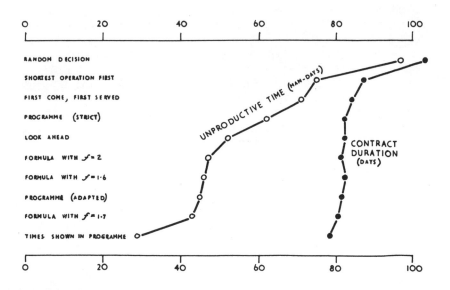

Fig. 6.48

6.9   BIDDING STRATEGY

6.9.1 Introduction

The majority of contracts in this country are obtained via competitive tendering. If a contractor is to be successful he must always be aware of what competitors are doing. If he knew the tender figures of his competitors he would be in a very advantageous position indeed. As he does not know this he must find ways of assessing what their tenders are likely to be to enable him to obtain contracts as profitably as possible, i.e. to enable him to maximise his profit. He must therefore use bidding strategy based on scientific methods. In order to apply the scientific method of statistics to tendering, the contractor needs to know the performance of his competitors in previous competitions. It is the practice of some architects and Local Authorities to inform contractors tendering, of the tenders submitted by all competitors either by opending tenders whilst contractors are present or by informing them after opening the tenders. If the contractor collects this information over a period of time he can learn quite a lot about his competitors.

6.9.2 Factors affecting the approach to the problem

When a contractor is invited to tender for work, he is not normally told who his competitors are. However, it is often possible to find out who is competing for a particular project. Even if the contractor does not know precisely who his competitors are, in certain circumstances he at least knows approximately how many competitors he has particularly if the "Code of Procedure for Selective Tendering" is being used. In the worst case he does not know who his competitors are nor how many there are. Each of these situations necessitates a slightly different approach.

6.9.3 1. When names of competing firms can be ascertained

When the contractor can ascertain who the competitors are, he can assess the optimum tender figure to give the maximum expected profit (expected profit being the average profit he would make if he used the same profit margin on a large number of contracts where the estimate of cost was identical and probability of being awarded the contracts remained the same), e.g. if the estimate of cost for a contract is £100000 and the contractor submits a tender figure of £110000, if he is awarded the contract he makes a profit of £10000, if he is not awarded the contract he gets nothing. If the probability of being awarded the contract is 0.4 the expected profit will be £10000 x 0.4 = £4000. Assume that information has been collected over a period of time with respect to a particular competitor and that these figures have been compared with the contractors estimated cost before profit is added as shown in Fig. 6.49.

6.9.3.1 *Frequency of ratios*

The information from Fig. 6.49 can be tabulated as shown in Fig. 6.50 which gives the frequency of occurrence of the various ratios.

6.9.3.2 *Probability of ratios*

From this table the probability of any particular ratio occurring can be found as shown in Fig. 6.51.

Contractors estimate	Competitors tender	Ratio of competitors Tender to contractors Estimate (T/E)
£120000	£126000	1.05
£150000	£159000	1.06
£160000	£166400	1.04
£140000	£148400	1.06
£130000	£139100	1.07
£110000	£111100	1.01
£160000	£177600	1.11
£140000	£151200	1.08
£100000 etc.	£107000 etc.	1.07 etc.

Fig. 6.49

T/E	Frequency
0.98	1
0.99	2
1.00	2
1.01	3
1.02	4
1.03	6
1.04	9
1.05	13
1.06	18
1.07	22
1.08	23
1.09	19
1.10	14
1.11	10
1.12	7
1.13	4
1.14	2
1.15	1
Total	160

Fig. 6.50

T/E	*Probability of occurrence*	*(frequency)* *(total)*
0.98	1/160	0.006
0.99	2/160	0.013
1.00	2/160	0.013
1.01	3/160	0.019
1.02	4/160	0.025
1.03	6/160	0.038
1.04	9/160	0.056
1.05	13/160	0.081
1.06	18/160	0.112
1.07	22/160	0.137
1.08	23/160	0.144
1.09	19/160	0.119
1.10	14/160	0.087
1.11	10/160	0.062
1.12	7/160	0.044
1.13	4/160	0.025
1.14	2/160	0.013
1.15	1/160	0.006
		1.000

Fig. 6.51

The contractor has now got a considerable amount of information about this particular competitor. He knows that 0.144 or 14.4% of his competitors tenders were 8% higher than his own estimate of cost, 0.6% were in fact 2% lower than his estimate of cost and 0.6% were 15% higher than his estimate.

6.9.3.3 *Cumulative probability distribution*
From Fig. 6.51 the cumulative probability distribution for this competitor can be found. The contractor will then know the probability of any tender being lower than this competitor, e.g. the probability of a tender of 0.98 or 98% of the estimate has a probability of being lower than this competitor of $1 - 0.006 = 0.994$. As the next tender/estimate ratio is 0.99 it is assumed that the probability of 0.994 applies for tenders from 0.98 to 0.989, and a tender less than 0.98 of the estimate has a probability of 1.00 of being lower than this competitor.

It is of course possible to carry out the analysis in a more sophisticated way than has been shown with much smaller intervals than the 0.01 or 1% intervals used in this analysis. Fig. 6.52 shows the cumulative probability distribution for this competitor.

T/E	Probability that tender will be less than competitor	
0.979		1.000
0.989	1 − 0.006	0.994
0.999	1 − 0.019	0.981
1.009	1 − 0.032	0.968
1.019	1 − 0.051	0.949
1.029	1 − 0.076	0.924
1.039	1 − 0.114	0.886
1.049	1 − 0.170	0.830
1.059	1 − 0.251	0.749
1.069	1 − 0.363	0.637
1.079	1 − 0.500	0.500
1.089	1 − 0.644	0.356
1.099	1 − 0.763	0.237
1.109	1 − 0.850	0.150
1.119	1 − 0.912	0.088
1.129	1 − 0.956	0.044
1.139	1 − 0.981	0.019
1.149	1 − 0.994	0.006
1.159	1 − 1.000	0.000

Fig. 6.52

6.9.3.4 Expected profit

Once the probability of a tender being less than that of the competitors is known, it is simple to calculate the expected profit for each ratio, e.g. if the contractor tenders using a 4.9% profit, i.e. using a ratio of 1.049 then the probability of success is 0.83. His expected profit is therefore 0.049 x estimate x 0.83 or $0.049E \times 0.83 = 0.041E$.

Figure 6.53 shows the expected profit for each ratio and from this it can be seen that the maximum expected profit is 0.442E or 4.42% using a ratio of 1.059, i.e. a profit margin of 5.9% of the estimated cost.

6.9.4 Competing against more than one known competitor

If there is more than one known competitor, the method of calculating the expected profit is very similar to that used for one competitor.

The result would be a number of cumulative probability distributions, the probability of being less than all competitors would be the product of the probabilities of being less than each individually.

If two such competitors are considered the result may be as shown in Fig. 6.54.

T/E	Expected profit
0.979	$1.000(0.979E - E) = -0.021E$
0.989	$0.994(0.989E - E) = -0.019E$
0.999	$0.981(0.999E - E) = -0.0010E$
1.009	$0.968(1.009E - E) = 0.0087E$
1.019	$0.949(1.019E - E) = 0.0180E$
1.029	$0.924(1.029E - E) = 0.0268E$
1.039	$0.886(1.039E - E) = 0.0346E$
1.049	$0.830(1.049E - E) = 0.0407E$
1.059	$0.749(1.059E - E) = 0.0442E$
1.069	$0.637(1.069E - E) = 0.0440E$
1.079	$0.500(1.079E - E) = 0.0395E$
1.089	$0.356(1.089E - E) = 0.0317E$
1.099	$0.237(1.099E - E) = 0.0235E$
1.109	$0.150(1.109E - E) = 0.0164E$
1.119	$0.088(1.119E - E) = 0.0105E$
1.129	$0.044(1.129E - E) = 0.0057E$
1.139	$0.019(1.139E - E) = 0.0026E$
1.149	$0.006(1.149E - E) = 0.0009E$
1.159	$0.000(1.159E - E) = 0.0000E$
	(to 4 decimal places)

Fig. 6.53

When the probability of tendering lower than all the known competitors has been calculated, the expected profit for each tender/estimate ratio can be worked out and the estimate showing the greatest expected profit can be selected as shown in Fig. 6.55

6.9.5 When names of competing firms cannot be ascertained
When the competitors are not known but the number is known a slight adjustment to the method is necessary. When using the "Code of Procedure for Selective Tendering" the maximum number of competitors will be known as this is laid down in the code. Usually this figure will be the nearest the contractor will get to the actual number of competitors.

In this situation the only difference in method is that the probability of submitting a lower tender is calculated against the average of all previous competitors. When this is ascertained the probability of a tender being lower than any number of competitors can be found as before, e.g. if a tender which includes a profit margin of 6.9%, i.e. a tender figure of 1.069E, has a probability of success of 0.600 against one competitor the probability against three such competitors would be $0.600 \times 0.600 \times 0.600$ or $0.63 = 0.216$.

Probability that tender will be less than

T/E	Competitor 1	Competitor 2	Competitors 1 and 2
0.979	1.000	1.000	1.000
0.989	0.994	0.990	0.984
0.999	0.981	0.975	0.956
1.009	0.968	0.961	0.930
1.019	0.949	0.941	0.893
1.029	0.924	0.918	0.848
1.039	0.886	0.878	0.778
1.049	0.830	0.824	0.684
1.059	0.749	0.746	0.559
1.069	0.637	0.636	0.405
1.079	0.500	0.502	0.251
1.089	0.356	0.359	0.128
1.099	0.237	0.238	0.056
1.109	0.150	0.155	0.023
1.119	0.088	0.100	0.009
1.129	0.044	0.046	0.002
1.139	0.019	0.025	0.000 (to 4 dec. places)
1.149	0.006	0.010	0.000 (to 4 dec. places)
1.159	0.000	0.005	0.000 (to 4 dec. places)

Fig. 6.54

6.9.6 Tendering against an unknown number of competitors

If statistics are to be used in this case an estimate of the number of competitors must be made. It is considered bad practice to ask unlimited numbers of contractors to tender for the work but if this is done and the number cannot be estimated, e.g. in open-tendering then statistics can be of very limited use.

6.9.7 Conclusions

It can be seen that by using statistics a contractor can gain an advantage over competitors who do not use analytical techniques in helping to determine profit margins on contracts. It must be appreciated that the main concern in this section has been to indicate the value of statistical methods for use in tendering. In practice other factors have to be taken into account besides the ones considered here. Indeed there are many other influencing factors in determining profit margins when tendering, examples being the state of the market and the economic climate at the time of tendering but this does not render statistical analysis irrelevant. Any method which assists in giving a more realistic appraisal of the situation should be taken into account.

T/E	Expected profit
0.979	$1.000(0.979E - E) = -0.021E$
0.989	$0.984(0.989E - E) = -0.0108E$
0.999	$0.956(0.999E - E) = -0.0010E$
1.009	$0.930(1.009E - E) = 0.0084E$
1.019	$0.893(1.019E - E) = 0.0170E$
1.029	$0.848(1.029E - E) = 0.0246E$
1.039	$0.778(1.039E - E) = 0.0303E$
1.049	$0.684(1.049E - E) = 0.0335E$
1.059	$0.559(1.059E - E) = 0.0330E$
1.069	$0.305(1.069E - E) = 0.021E$
1.079	$0.251(1.079E - E) = 0.0198E$
1.089	$0.128(1.089E - E) = 0.0114E$
1.099	$0.056(1.099E - E) = 0.0055E$
1.109	$0.023(1.109E - E) = 0.0025E$
1.119	$0.009(1.119E - E) = 0.0011E$
1.129	$0.002(1.129E - E) = 0.0002E$
1.139	$0.000(1.139E - E) = 0.000E$
1.149	$0.000(1.149E - E) = 0.000E$
1.159	$0.000(1.159E - E) = 0.000E$

Fig. 6.55

6.10 USE OF COMPUTERS

Standard software packages are available for solving many operational research problems. Typical examples being packages to solve linear programming problems, transportation problems, problems using simulation and of course Critical Path Method and its extensions in the form of resource allocation and cost optimisation.

Probably the greatest advantage of using the computer is its speed which enables problems, which would take months by hand to be done in a small fraction of the time. Indeed many problems could not be solved at all within a practical time without the use of a computer. Many simulation problems fall into this category.

Wherever possible it is essential that an integrated data processing system is used as in this case much of the information required to solve a problem will probably be readily available in store, the information being collected as part of a total system.

Reference was made at the beginning of this book to integrated management information systems and in large firms a computer would be essential if an integrated system is to be run efficiently.

Bibliography

1: CONSTRUCTION PLANNING AND CONTROL

Introduction to building Management. R. E. CALVERT George Newnes Ltd.
The Programming of House Building. Building Research Station Digest No. 91.
H.M.S.O.
Production Planning Applied to Building. R. J. HOLLINS George Godwin for
The Builder Ltd.
Site Records for Builders. H.M.S.O.
Programming & Progressing. H.M.S.O.

2: THE CRITICAL PATH METHOD OF PLANNING

Critical Path Methods in Construction Practice. J. M. ANTILL & R. W. WOODHEAD
Wiley.

BUILDING RESEARCH STATION CURRENT PAPERS
1962/3, 'The Critical Path Method'. J. F. NUTTALL, *The Builder*, vol. 204.
1964/3, 'Critical Path Methods Applied to Building Site Control'.
J. F. NUTTALL & E. E. AMOS, *The Builder*, vol. 207, 1964.
1966/25, 'Resource scheduling: coping with labour problems as they
occur on site'. J. F. NUTTALL, *The Builder*, vol. 210, 1966.
1966/29. 'Network diagrams: some notes on alternative presentations'.
R. E. JEANES, *Building*, 1966, vol. 210.
1966/39. 'Computerising resource needs'. J. R. BRITTEN. *Building*, 1967,
17th February.
1966/39A. 'Computerising resource needs'. Appendix. J. R. BRITTEN,
Building, 1967, 17th February.
1966/41. 'The advantages of time-scaled networks and planning frames'.
J. R. BRITTEN, *National Builder*, 1967, vol. 48.
1965/34. 'The control of repetitive construction'. J. F. NUTTALL;

3: WORK STUDY

Glossary of Terms in Work Study BS 3138, 1959 British Standards Institution.
Outline of Work Study, Parts 1, 2, 3, The British Institute of Management.
Work Study Applied to Building. R. GEARY, *The Builder* Ltd.
'Some Principles of Production Control of Building Work; The application of
Queuing Theory and of Simulation Techniques'. J. F. NUTTALL, *Journal of
Industrial Economics,* Nov. 1961.
Project Planning & Control in the Construction Industry. G. J. TURNER &
K. R. J. ELLIOT, Cassell.
Work Study. R. M. CURRIE, British Institute of Management.

4: STATISTICS

An Outline of Statistics, 7th Edition. SAMUEL HAYS, Longmans, Green.
Elementary Statistical Methods. E. C. RHODES, Routledge, Kegan & Paul.
What do Statistics Show? F. R. OLIVER, Hodder & Stoughton Ltd.
Facts from Figures. M. J. MORONEY, Penguin Books.
Use & Abuse of Statistics. W. J. REICHMAN, Pelican Books.
Business Charts. T. G. ROSE, Pitman.
Principles of Statistical Techniques. P. G. MOORE, Cambridge University Press.
Concrete & Statistics. J. D. McINTOSH, C. R. Books Ltd.
Statistics & the Manager. P. G. MOORE, Macdonald.

5: BUDGETARY AND COST CONTROL

Builders Cost Control, Bonusing and Accounts. F. BRANDWOOD. Gee & Co.
Introduction to Budgetary Control, Standard Costing, Material Control & Production Control. Institute of Cost & Works Accountants, Gee & Co.
Budgetary Control & Standard Costs. J. A. SCOTT, Pitman.

6: OPERATIONAL RESEARCH

Operational Research and Quantitative Economics —An elementary introduction. HENRY THEIL, JOHN C. G. BOOT and TEUN KLOEK. McGraw Hill.
Construction and Professional Management — An introduction. HARRY RUBEY and WALKER W. MILNER. Macmillan.
A Guide to Operational Research. ERIC DUCKWORTH. University paperbacks.
Operational Research for Management. M. J. SARGEAUNT. Heinemann.
Research Problems in Production routing and scheduling. A. J. ROWE and J. R. JACKSON. Journal of Industrial Economics. 1956 — 7(3) pp. 116—21.

7: GENERAL

Systems Analysis and Design Applied to the Construction Industry. R. OXLEY and J. POSKITT. Building Technology and Management. December 1969, and *The Illustrated Carpenter & Builder.* March 19th 1971.

Index